Great Trees of India

Ruskin Bond is known for his signature simplistic and witty writing style. He is the author of several bestselling short stories, novellas, collections, essays and children's books; and has contributed a number of poems and articles to various magazines and anthologies. At the age of 23, he won the prestigious John Llewellyn Rhys Prize for his first novel, *The Room on the Roof*. He was also the recipient of the Padma Shri in 1999, Lifetime Achievement Award by the Delhi Government in 2012 and the Padma Bhushan in 2014.

Born in 1934, Ruskin Bond grew up in Jamnagar, Shimla, New Delhi and Dehradun. Apart from three years in the UK, he has spent all his life in India and now lives in Landour, Mussoorie, with his adopted family.

RUSKIN BOND
Great Trees of India

RUPA

Published by
Rupa Publications India Pvt. Ltd 2025
7/16, Ansari Road, Daryaganj
New Delhi 110002

Sales centres:
Bengaluru Chennai
Hyderabad Jaipur Kathmandu
Kolkata Mumbai Prayagraj

Copyright © Ruskin Bond 2025

All rights reserved.
This is a work of fiction. Names, characters, places and incidents are either the product of the author's imagination or are used fictitiously and any resemblance to any actual person, living or dead, events or locales is entirely coincidental.

No part of this publication may be reproduced, transmitted, or stored in a retrieval system, in any form or by any means, electronic, mechanical, photocopying, recording or otherwise, without the prior permission of the publisher.

P-ISBN: 978-93-6156-083-5
E-ISBN: 978-93-6156-916-6

First impression 2025

10 9 8 7 6 5 4 3 2 1

The moral right of the author has been asserted.

Printed in India

This book is sold subject to the condition that it shall not, by way of trade or otherwise, be lent, resold, hired out, or otherwise circulated, without the publisher's prior consent, in any form of binding or cover other than that in which it is published.

CONTENTS

Introduction vii

1. Growing Up with Trees 1
2. The Trees Are My Brothers 4
3. My Father's Trees in Dehra 7
4. Music in the Trees 18
5. The Rhododendrons 22
6. Death of the Trees 24
7. A Good Place for Trees 27
8. Gentle Shade by Day 32
9. The Cherry Tree 34
10. Under the Deodars 39
11. The Friendly Oaks 40
12. Great Trees of Garhwal 41
13. The Prospect of Flowers 46
14. The Friendly Banyan 53
15. Angry River 54
16. The Banyan Tree 80
17. The Coral Tree 84
18. Rhododendrons in the Mist 87
19. Among the Maples and Oaks 98
20. Guests Who Fly in from the Forest 101

21. Great Spirits of the Trees 104
22. Wild Flowers near a Mountain Stream 108
23. The School among the Pines 111
24. In the Garden of My Dreams 133

INTRODUCTION

Throughout life, the trees here in India have been my constant companions. From hours whiled away in childhood in the recesses of their branches, to moments years later spent looking at the ones that grew outside my window, time has only strengthened my love for them. Their giving soul, their sturdy character, their unending variety, the stories associated with them—all are aspects that remind me of the people of this country and why they are so close to my heart.

Great Trees of India contains a selection of my writings about these simple, fortunately common, marvels of nature. Stories like 'Growing Up with Trees' and 'My Father's Trees in Dehra' are about trees from my childhood and the strong memories of my family that are deeply intertwined with them. 'Music in the Trees' and 'Guests Who Fly in from the Forest' are about the vibrant ecosystem of insects, birds and wildlife that trees house, which are just as lovely to me as the trees themselves. On the other hand are 'Death of the Trees' and 'Angry River', stories in which trees, like all living things, fall, due to either human intervention or just the passage of time. Lastly, there are short pieces on some of my favourite varieties of trees, such as 'Under the Deodars', 'The Friendly Oaks' and 'The Rhododendrons', the last mentioned also forming an eerie backdrop to one of the darker stories in the collection, 'Rhododendrons in the Mist'.

In all their colours, shapes and forms, the trees and plants of India are gifts that have brought endless beauty and joy to my life. I hope the stories here will inspire you to look out your window, step into your garden, and remember why trees have always been such an essential part of our existence.

Ruskin Bond

GROWING UP WITH TREES

Dehradun was a place for trees, and Grandfather's house was surrounded by several kinds—peepul, neem, mango, jackfruit and papaya. There was also an ancient banyan tree. I grew up amongst these trees, and some of them, planted by Grandfather, grew with me.

There were two types of trees that were of special interest to a boy—trees that were good for climbing, and trees that provided fruit.

The jackfruit tree was both these things. The fruit itself—the largest in the world—grew only on the trunk and main branches. I did not care much for the fruit, although cooked as a vegetable, it made a good curry. But the tree was large and leafy and easy to climb. It was a very dark tree and if I hid in it, I could not easily be seen from below. In a hole in the tree trunk I kept various banned items—a catapult, some lurid comics, and a large stock of chewing-gum. Perhaps they are still there, because I forgot to collect them when we finally went away.

The banyan tree grew behind the house. Its spreading branches, which hung to the ground and took root again, formed a number of twisting passageways which gave me endless pleasure. The tree was older than the house, older than my grandparents, as old as Dehra. I could hide myself in its branches, behind thick green leaves, and spy on the world below. I could read in it, too, propped up against the bole of the tree, with *Treasure Island* or the *Jungle Book*s or comics like *Wizard* or *Hotspur* which, unlike the forbidden *Superman* and others like him, were full of clean-cut schoolboy heroes.

The banyan tree was a world in itself, populated with small beasts and large insects. While the leaves were still pink and tender, they would be visited by the delicate Map Butterfly, who committed her eggs to their care. The 'honey' on the leaves—an edible smear—also attracted the little striped squirrels, who soon grew used to my presence and became quite bold. Red-headed parakeets swarmed about the tree early in the mornings.

But the banyan really came to life during the monsoon, when the branches were thick with scarlet figs. These berries were not fit for human consumption, but the many birds that gathered in the tree—gossipy Rosy Pastors, quarrelsome Mynas, cheerful Bulbuls and Coppersmiths, and sometimes a raucous bullying crow—feasted on them. And when night fell, and the birds were resting, the dark Flying Foxes flapped heavily about the tree, chewing and munching as they clambered over the branches.

Among nocturnal visitors to the jackfruit and banyan trees was the Brainfever bird, whose real name is the Hawk-Cuckoo. 'Brainfever, brainfever!' it seems to call, and this shrill, nagging cry will keep the soundest of sleepers awake on a hot summer's night.

The British called it the Brainfever bird, but there are other names for it. The Marathas called it 'Paos-ala,' which means 'Rain is coming!' Perhaps Grandfather's interpretation of its call was the best. According to him, when the bird was tuning up for its main concert, it seemed to say: 'Oh dear, oh dear! How very hot it's getting! We feel it... we feel it... WE FEEL IT!'

Yes, the banyan tree was a noisy place during the rains. If the Brainfever bird made music by night, the crickets and cicadas orchestrated during the day. As musicians, the cicadas were in a class by themselves. All through the hot weather their chorus rang through the garden, while a shower of rain, far from damping their spirits, only roused them to a greater vocal effort.

The tree-crickets were a band of willing artistes who commenced their performance at almost any time of the day but preferably in the evenings. Delicate pale green creatures with transparent green wings, they were hard to find amongst the lush monsoon foliage; but once located, a tap on the leaf or bush on which they sat would put an immediate end to the performance.

At the height of the monsoon, the banyan tree was like an orchestra-pit with the musicians constantly turning up. Birds, insects and squirrels expressed their joy at the end of the hot weather and the cool quenching relief of the rains.

A flute in my hands, I would try adding my shrill piping to theirs. But they thought poorly of my musical ability, for, whenever I played on the flute, the birds and insects would subside into a pained and puzzled silence.

THE TREES ARE MY BROTHERS

It's good to know that my old friend the jackfruit is finally coming into its own. Apparently it is now much in demand in the West, a fashionable substitute for meat, being used as filling for burgers, sandwiches, pies etc., with one enthusiast even calling it 'mutton hanging from a tree'.

Here in India we have always appreciated a good jackfruit curry, or even better, a jackfruit pickle. I'm a pickle fiend myself, and among the twenty different pickles on my sideboard there is always a jar of jackfruit pickle; that's why I call it an old friend. But I had no idea it tasted like mutton. The seed and the pulp have their own individual flavour. As it grows on a tree we call it a fruit, but we cook it as though it were a vegetable. And if, to some, it tastes like mutton, then perhaps some meat-eaters will become vegetarians. On the other hand, some vegetarians might not care for its meaty flavour!

When I was a boy, we had an old jackfruit tree growing beside the side verandah. I spent a lot of time in the trees surrounding my grandmother's bungalow, and this one was easy to climb. The others included several guava and lichi trees, lemons and grapefruits, and of course a couple of mango trees—but these last were difficult to climb.

'Why do you spend so much time in the trees?' complained my grandmother. 'Why not do something useful for a change?'

'The trees are my brothers,' I would say, 'I like to play with them.'

And I still think of them as my brothers, although I can no longer climb trees or play in them. But I still think of them as human beings possessed of individuality and charm. Just as no two humans are exactly alike (unless they happen to be twins), so no two trees are the same. Like humans they grow from seed. They develop branches as arms and leaves like flowing hair. We give birth to children, they give birth to fruits and flowers. We shelter our young, they shelter the small creatures of the forest.

However, unlike us, they spring from the soul, from the land—the very land that gives us food and pasture and protection; the land that we so casually take for granted, preferring to build upon it rather than grow upon it. Where will our cattle graze when the last green spaces have gone?

'No problem,' says a young friend. 'We can always import our milk.'

The other day I came across an old book that had been on my shelf for many years. *Farmer's Glory* by A.G. Street, written several decades ago. In his epilogue he writes:

> It is perhaps nothing to boast about, but there is little doubt that the present prosperity of British farming is mainly due to one man, who is now dead. His name was Adolf Hitler. There is no disputing that it was the fear of famine during the early 1940s which taught the British nation that despite all man's cleverness and inventions, when real danger comes an island, people must turn for succour to the only permanent asset they possess, the land of their own country. It has never, and will never, let them down; always provided they realize and obey this eternal truth—that to make the land serve man, man must first be content to serve the land.

Surely it is this love of the land and willingness to serve it that is at the heart of true patriotism. The patriotic songs and speeches that we hear from time to time are fine for stirring up the emotions, but it is really the connect between ourselves and the '*do bigha zameen*' on which we grow our fruit and grain that emboldens us to protect it.

I think I am correct in saying that most of our jawans, the young men who join the solid ranks of the Indian Army, come from rural backgrounds; some from the hills, some from the vast plains and hinterland of our country. They know the value of the land. They have grown up in villages and have worked with their families in the rice fields, or sugar cane plantations, or mango groves, or wheat or corn or mustard or fields of an infinite variety of crops. More than city folk, they know the value of the land—its true worth in terms of either prosperity or poverty. And so they are ready to defend it, to fight for it against all corners. The best soldiers come from the soil that they and their forefathers have tilled.

So let us protect the land—not just from the intruder or the enemy, but also from those who would turn the field or the forest into one more concrete jungle.

Of course there are those who prefer concrete jungles. Like my young friend who wants to live in a Smart City and never mind the cities that are no longer smart. My advice to him (unheeded of course) is to go back to his roots, create a smart little village, and plant jackfruit trees!

MY FATHER'S TREES IN DEHRA

Our trees still grow in Dehra. This is one part of the world where trees are a match for man. An old peepul may be cut down to make way for a new building; two peepul trees will sprout from the walls of the building. In Dehra the air is moist, the soil hospitable to seeds and probing roots. The valley of Dehradun lies between the first range of the Himalayas and the smaller but older Siwalik range. Dehra is an old town, but it was not in the reign of Rajput princes or Mogul kings that it really grew and flourished; it acquired a certain size and importance with the coming of British and Anglo-Indian settlers. The English have an affinity with trees, and in the rolling hills of Dehra they discovered a retreat, which, in spite of snakes and mosquitoes, reminded them, just a little bit, of England's green and pleasant land.

The mountains to the north are austere and inhospitable; the plains to the south are flat, dry and dusty. But Dehra is green. I look out of the train window at daybreak to see the sal and shisham trees sweep by majestically, while trailing vines and great clumps of bamboo give the forest a darkness and density that add to its mystery. There are still a few tigers in these forests; only a few, and perhaps they will survive, to stalk the spotted deer and drink at forest pools.

I grew up in Dehra. My grandfather built a bungalow on the outskirts of the town at the turn of the century. The house was sold a few years after Independence. No one knows me now in Dehra, for it is over twenty years since I left the place, and my boyhood friends are scattered and lost. And although the India of Kim is no more, and the Grand Trunk Road is now a procession of trucks instead of a slow-moving caravan of horses and camels, India is still a country in which people are easily lost and quickly forgotten.

From the station I can take either a taxi or a snappy little scooter rickshaw (Dehra had neither before 1950), but, because I am on an unashamedly sentimental pilgrimage, I take a tonga, drawn by a lean, listless pony, and driven by a tubercular old Muslim in a shabby green waistcoat. Only two or three tongas stand outside the station. There were always twenty or thirty here in the 1940s when I came home from boarding school to be met at the station by my grandfather; but the days of the tonga are nearly over, and in many ways this is a good thing, because most tonga ponies are overworked and underfed. Its wheels squeaking from lack of oil and its seat slipping out from under me, the tonga drags me through the bazaars of Dehra. A couple of miles of this slow, funereal pace makes me impatient to use my own legs, and I dismiss the tonga when we get to the small Dilaram Bazaar.

It is a good place from which to start walking.

The Dilaram Bazaar has not changed very much. The shops are run by a new generation of bakers, barbers and banias, but professions have not changed. The cobblers belong to the lower castes, the bakers are Muslims, the tailors are Sikhs. Boys still fly kites from the flat rooftops, and women wash clothes on the canal steps. The canal comes down from Rajpur and goes underground here, to emerge about a mile away.

I have to walk only a furlong to reach my grandfather's house. The road is lined with eucalyptus, jacaranda and laburnum trees. In the compounds there are small groves of mangoes, litchis and papayas. The poinsettia thrusts its scarlet leaves over garden walls. Every verandah has its bougainvillea creeper, every garden its bed of marigolds. Potted palms, those symbols of Victorian snobbery, are popular with Indian housewives. There are a few houses, but most of the bungalows were built by 'old India hands' on their retirement from the army, the police or the railways. Most of the present owners are Indian businessmen or government officials.

I am standing outside my grandfather's house. The wall has been raised, and the wicket gate has disappeared; I cannot get a clear view of the house and garden. The nameplate identifies the owner as Major General Saigal; the house has had more than one owner since my grandparents sold it in 1949.

On the other side of the road there is an orchard of litchi trees. This is not the season for fruit, and there is no one looking after the garden. By taking a little path that goes through the orchard, I reach higher ground and gain a better view of our old house.

Grandfather built the house with granite rocks taken from the foothills. It shows no sign of age. The lawn has disappeared; but the big jackfruit tree, giving shade to the side verandah, is still there. In this tree I spent my afternoons, absorbed in my Magnets, Champions and Hotspurs, while sticky mango juice trickled down my chin. (One could not eat the jackfruit unless it was cooked into a vegetable curry.) There was a hole in the bole of the tree in which I kept my pocket knife, top, catapult and any badges or buttons that could be saved from my father's RAF tunics when he came home on leave. There was also an Iron Cross, a relic of the First World War, given to me by my grandfather. I have managed to keep the Iron Cross; but what did I do with my top and catapult? Memory fails me. Possibly they are still in the hole in the jackfruit tree; I must have forgotten to collect them when we went away after my father's death. I am seized by a whimsical urge to walk in at the gate, climb into the branches of the jackfruit tree and recover my lost possessions. What would the present owner, the major general (retired), have to say if I politely asked permission to look for a catapult left behind more than twenty years ago?

An old man is coming down the path through the litchi trees. He is not a major general but a poor street vendor. He carries a small tin trunk on his head, and walks very slowly. When he sees me, he stops and asks me if I will buy something. I can think of nothing I need, but the old man looks so tired, so very old, that I am afraid he will collapse if he moves any further along the path without resting. So I ask him to show me his wares. He cannot get the box off his head by himself, but together we manage to set it down in the shade, and the old man insists on spreading its entire contents on the grass; bangles, combs, shoelaces, safety pins, cheap stationery, buttons, pomades, elastic and scores of other household necessities.

When I refuse buttons because there is no one to sew them on for me, he plies me with safety pins. I say no; but as he moves from one article to another, his querulous, persuasive voice slowly wears down my resistance, and I end up buying envelopes, a letter pad (pink roses on bright blue paper), a one-rupee fountain pen guaranteed to leak and several yards of elastic. I have no idea what I will do with the elastic, but the old man convinces me that I cannot live without it.

Exhausted by the effort of selling me a lot of things I obviously do not want, he closes his eyes and leans back against the trunk of a litchi tree. For a moment I feel rather nervous. Is he going to die sitting here beside me? He sinks to his haunches and puts his chin on his hands. He only wants to talk.

'I am very tired, *huzoor*,' he says. 'Please do not mind if I sit here for a while.'

'Rest for as long as you like,' I say. 'That's a heavy load you've been carrying.'

He comes to life at the chance of a conversation and says, 'When I was a young man, it was nothing. I could carry my box up from Rajpur to Mussoorie by the bridle path—seven steep miles! But now I find it difficult to cover the distance from the station to Dilaram Bazaar.'

'Naturally. You are quite old.'

'I am seventy, sahib.'

'You look very fit for your age.' I say this to please him; he looks frail and brittle. 'Isn't there someone to help you?' I ask.

'I had a servant boy last month, but he stole my earnings and ran off to Delhi. I wish my son was alive—he would not have permitted me to work like a mule for a living—but he was killed in the riots in 1947.'

'Have you no other relatives?'

'I have outlived them all. That is the curse of a healthy life. Your friends, your loved ones, all go before you, and in the end you are left alone. But I must go too, before long. The road to the bazaar seems to grow longer every day. The stones are harder. The sun is hotter in the summer, and the wind much colder in the winter. Even some of the trees that were there in my youth have grown old and have died. I have outlived the trees.'

He has outlived the trees. He is like an old tree himself, gnarled and twisted. I have the feeling that if he falls asleep in the orchard, he will strike root here, sending out crooked branches. I can imagine a small bent tree wearing a black waistcoat; a living scarecrow.

He closes his eyes again, but goes on talking.

'The English memsahibs would buy great quantities of elastic. Today it is ribbons and bangles for the girls, and combs for the boys. But I do not make much money. Not because I cannot walk very far. How many houses do I reach in a day?

Ten, fifteen. But twenty years ago I could visit more than fifty houses. That makes a difference.'

'Have you always been here?'

'Most of my life, huzoor. I was here before they built the motor road to Mussoorie. I was here when the sahibs had their own carriages and ponies and the memsahibs their own rickshaws. I was here before there were any cinemas. I was here when the Prince of Wales came to Dehradun... Oh, I have been here a long time, huzoor. I was here when that house was built,' he says pointing with his chin towards my grandfather's house. 'Fifty, sixty years ago it must have been. I cannot remember exactly. What is ten years when you have lived seventy? But it was a tall, red-bearded sahib who built that house. He kept many creatures as pets. A *kachwa* (turtle) was one of them. And there was a python, which crawled into my box one day and gave me a terrible fright. The sahib used to keep it hanging from his shoulders, like a garland. His wife, the burra mem, always bought a lot from me—lots of elastic. And there were sons, one a teacher, another in the air force, and there were always children in the house. Beautiful children. But they went away many years ago. Everyone has gone away.'

I do not tell him that I am one of the 'beautiful children'. I doubt if he will believe me. His memories are of another age, another place, and for him there are no strong bridges into the present.

'But others have come,' I say.

'True, and that is as it should be. That is not my complaint. My complaint—should God be listening—is that I have been left behind.'

He gets slowly to his feet and stands over his shabby tin box, gazing down at it with a mix of disdain and affection. I help him to lift and balance it on the flattened cloth on his head. He does not have the energy to turn and make a salutation of any kind; but, setting his sights on the distant hills, he walks down the path with steps that are shaky and slow but still wonderfully straight.

I wonder how much longer he will live. Perhaps a year or two, perhaps a week, perhaps an hour. It will be an end of living, but it will not be death. He is too old for death; he can only sleep; he can only fall gently, like an old, crumpled brown leaf.

I leave the orchard. The bend in the road hides my grandfather's house. I reach the canal again. It emerges from under a small culvert, where ferns and maidenhair grow in the shade. The water, coming from a stream in the foothills, rushes along with a familiar sound; it does not lose its momentum until the canal has left the gently sloping streets of the town.

There are new buildings on this road, but the small police station is housed in the same old lime-washed bungalow. A couple of off-duty policemen, partly uniformed but with their pyjamas on, stroll about.

I cannot forget this little police station. Nothing very exciting ever happened in its vicinity until, in 1947, communal riots broke out in Dehra. Then, bodies were regularly fished out of the canal and dumped on a growing pile in the station compound. I was only a boy, but when I looked over the wall at that pile of corpses, there was no one who paid any attention to me. They were too busy to send me away. At the same time they knew that I was perfectly safe; while Hindus and Muslims were at each other's throats, a white boy could walk the streets in safety. No one was any longer interested in the Europeans.

The people of Dehra are not violent by nature, and the town has no history of communal discord. But when refugees from the partitioned Punjab poured into Dehra in thousands, the atmosphere became charged with tension. These refugees, many of them Sikhs, had lost their homes and livelihoods; many had seen their loved ones butchered. They were in a fierce and vengeful frame of mind. The calm, sleepy atmosphere of Dehra was shattered during two months of looting and murder. The Muslims who could get away, fled. The poorer members of the community remained in a refugee camp until the holocaust was over; then they returned to their former occupations, frightened and deeply mistrustful. The old boxman was one of them.

I cross the canal and take the road that will lead me to the riverbed. This was one of my father's favourite walks. He, too, was a walking man. Often, when he was home on leave, he would say, 'Ruskin, let's go for a walk,' and we would slip off together and walk down to the riverbed or into the sugar cane fields or across the railway lines and into the jungle.

On one of those walks (this was before Independence), I remember him saying, 'After the war is over, we'll be going to England. Would you like that?'

'I don't know,' I said. 'Can't we stay in India?'

'It won't be ours any more.'

'Has it always been ours?' I asked.

'For a long time,' he said, 'over two hundred years. But we have to give it back now.'

'Give it back to whom?' I asked. I was only nine.

'To the Indians,' said my father.

The only Indians I had known till then were my ayah and the cook and the gardener and their children, and I could not imagine them wanting to be rid of us. The only other Indian who came to the house was Dr Ghose, and it was frequently said of him that he was more English than the English. I could understand my father better when he said, 'After the war, there'll be a job for me in England. There'll be nothing for me here.'

The war had at first been a distant event; but somehow it kept coming closer. My aunt, who lived in London with her two children, was killed with them during an air raid; then my father's younger brother died of dysentery on the long walk out from Burma. Both these tragic events depressed my father. Never in good health (he had been prone to attacks of malaria), he looked more worn and wasted every time he came home. His personal life was far from being happy, as he and my mother had separated, she to marry again. I think he looked forward a great deal to the days he spent with me; far more than I could have realized at the time. I was someone to come back to; someone for whom things could be planned; someone who could learn from him.

Dehra suited him. He was always happy when he was among trees, and this happiness communicated itself to me. I felt like drawing close to him. I remember sitting beside him on the verandah steps when I noticed the tendril of a creeping vine that was trailing near my feet. As we sat there, doing nothing in particular—in the best gardens, time has no meaning—I found that the tendril was moving almost imperceptibly away from me and towards my father. Twenty minutes later it had crossed the verandah steps and was touching his feet. This, in India, is the sweetest of salutations.

There is probably a scientific explanation for the plant's behaviour—something to do with the light and warmth on the verandah steps—but I like to think that its movements were motivated simply by an affection for my father. Sometimes, when I sat alone beneath a tree, I felt a little lonely or lost. As soon as my father rejoined me, the atmosphere lightened; the tree itself became more friendly.

Most of the fruit trees round the house were planted by father; but he was not content with planting trees in the garden. On rainy days we would walk beyond the riverbed, armed with cuttings and saplings, and then we would amble through the jungle, planting flowering shrubs between the sal and shisham trees.

'But no one ever comes here,' I protested the first time. 'Who is going to see them?'

'Some day,' he said, 'someone may come this way... If people keep cutting trees instead of planting them, there'll soon be no forests left at all, and the world will be just one vast desert.' The prospect of a world without trees became a sort of nightmare for me (and one reason why I shall never want to live on the treeless moon), and I assisted my father in his tree planting with great enthusiasm.

'One day the trees will move again,' he said. 'They've been standing still for thousands of years. There was a time when they could walk about like people, but someone cast a spell on them and rooted them to one place. But they're always trying to move—see how they reach out with their arms!'

We found an island, a small rocky island in the middle of a dry riverbed. It was one of those riverbeds, so common in the foothills, that are completely dry in the summer but flooded during the monsoon rains. The rains had just begun, and the stream could still be crossed on foot, when we set out with a number of tamarind, laburnum and coral tree saplings and cuttings. We spent the day planting them on the island, then ate our lunch there, in the shelter of a wild plum.

My father went away soon after that tree planting. Three months later, in Calcutta, he died.

I was sent to boarding school. My grandparents sold the house and left Dehra. After school, I went to England. The years passed, my grandparents died, and when I returned to India, I was the only member of the family in the country.

And now I am in Dehra again, on the road to the riverbed. The houses with their trim gardens are soon behind me, and I am walking through fields of flowering mustard, which make a carpet of yellow blossom stretching away towards the jungle and the foothills.

The riverbed is dry at this time of the year. A herd of skinny cattle graze on the short brown grass at the edge of the jungle. The sal trees have been thinned out. Could our trees have survived? Will our island be there, or has some flash flood during a heavy monsoon washed it away completely?

As I look across the dry watercourse, my eye is caught by the spectacular red plumes of the coral blossom. In contrast with the dry, rocky riverbed, the little island is a green oasis. I walk across to the trees and notice that a number of parrots have come to live in them. A koel challenges me with a rising *who-are-you, who-are you.*

But the trees seem to know me. They whisper among themselves and beckon me nearer. And looking around, I find that other trees and wild plants and grasses have sprung up under the protection of the trees we planted.

They have multiplied. They are moving. In this small forgotten corner of the world, my father's dreams are coming true, and the trees are moving again.

MUSIC IN THE TREES

In India, the monsoon is the season when our insect orchestra is at its best. It is true that the shrill music of the cicada is heard throughout the hot weather; but theirs is a prelude to the great concert that comes into full play once the rainy season begins. When the monsoon with its magic touch brings life and greenness to rock and earth and tree, the whole air seems to come alive with the music of insects. Grasshoppers shrill in the bushes, crickets chirp from under stones, and in the water-laden fields there are hundreds of minor artists providing a medley of sounds.

Amongst our more vocal and better-known insect musicians are those that dwell in trees, the cicadas and the crickets. As musicians, the cicadas are in a class by themselves. Most of the species in India are forest dwellers, but there are some who inhabit the open country in the plains. All through the hot weather their chorus rings through the jungle, while a shower of rain, far from damping their spirits, only rouses them to a deafening, combined effort. The ancient Greeks knew the cicada well. They appreciated his music so much that they kept him captive in a cage to sing. Only the males were chosen, for the females, like most insect musicians, were completely dumb. This moved one of the Greek poets to exclaim, 'Happy are the cicadas, for they have voiceless wives.'

The cicada's sound-producing organs are amongst the most remarkable in the animal kingdom. The underside of his body carries a pair of flaps, each of which covers an oval membrane, which looks like the head of a drum. These are set into motion by a great pair of muscles attached to them from within the body, and the sound is produced by their vibration. The whole abdomen, which is practically hollow, helps to increase or diminish the sound.

Simple, isn't it? To be truthful, I find it extremely complicated, and am able to describe the process only by consulting the notes of S.H. Prater, one-time curator of the Bombay Natural History Society.

Let it be added that the female carries these structures in a modified form, but as she has no muscles to bring them into play, she is unable to use them. This is why she must remain silent while her spouse shrieks away. I would change the line from that Greek poet (Xenarchos, I think) and say instead: 'Pity the female cicadas, for they have singing husbands!'

The object of the cicadas' mirthful music is a mystery. It may attract the opposite sex, or it may be just a diversion of the male. Or perhaps he sings because he is happy.

The tree crickets are a band of willing artists who commence their performance as soon as it is dusk. Their sounds are familiar, but the crickets are seldom seen. If one of them enters the house and treats us to a solo, the sound is so surprisingly loud that we can hardly believe it is being produced by so small a creature.

The common Indian tree cricket is a delicate pale-green little creature with hazy, transparent green wings. In full song he holds his wings outspread over his back. They vibrate so rapidly that they are but a blurred outline. A tap on the bush or leaf on which he sits will put an immediate stop to his performance. His music ceases, and he lowers his wings and folds them flat on his back. The grasshopper makes his music by rubbing his legs against his forewing.

I won't go into detail over how the cricket produces its music, except to say that its louder notes are produced by a rapid vibration of the wings, the right wing usually working over the left, and the edge of one acting on the file of the other to produce a shrill, long-sustained note.

One of the best-known crickets is a large black fellow who lives underground and rarely comes out by day, except when the rains flood him out of his burrow. But when night falls, he sits on his doorstep and pours out his soul in a strident song. This cricket's name is as impressive as his sound—*Brachytrypes portentosus.*

The mole cricket is a genius by itself. Mole crickets are tillers of the soil. They use their powerful forelimbs for shovelling up the earth and their hard heads for butting into it. Notwithstanding its earthy occupations, the mole cricket is sometimes moved to creating music. But as he repeats his note—a solemn deep-toned chirp, about a hundred times a minute—the performance can be monotonous.

In India, the cone-headed katydids are probably the most notable performers. Katydids are trim, slender grasshopper-like insects, much in evidence in the fresh green grass of the monsoon. In the fields, the loud, shrill notes of the males may be heard both by day and by night. Sometimes one of them comes into a house and treats its occupants to a sudden outburst of high-pitched fiddling. His song rises in pitch as the performer warms to his work. In a room it can be quite deafening, and the sound is always difficult to locate—it seems to come from everywhere.

Finally we come to the tree crickets, a band of willing artistes who commence their performance at dusk. Their sounds are familiar, but it is difficult to see the musicians. Presumably the males sing in order to attract their more silent females. The music advertises the presence of the male; just as in other creatures, it is the colour or smell that does the job. And if music be the food of love, play on, cicada!

Why are grasshoppers and crickets such persistent little singers? Do they really sing to charm and attract the females, or is their song the voice of mirth? A curious habit has been noticed among certain tree crickets, which may offer a clue to the mystery. Sometimes, as a male sings, a female steals up to him from behind. The male ceases his music. He sits quite still with his wings uplifted. The female noses about his back and soon discovers the object of her search—a deep cavity situated just behind the base of his wings. This cavity contains a clear liquid which she eagerly laps up. Well, even the human male seeks to please his sweetheart with the offer of chocolates.

It is supposed that in this instance, the lady is attracted rather by the sweets the male has to offer than by his music. But the music advertises his whereabouts. She hears his sound and knows that he has a sweet nectar to offer her and comes after it. If the artful luring of the male sometimes results in mating, we see the real reason for the male possessing his musical instruments, and understand his urge to play them so continuously. After all, the luring of the female with music and sweets is even practised by human beings. It may not always succeed in its purpose. Sometimes, as with the crickets, the female accepts the gifts so generously offered—and then takes her leave!

THE RHODODENDRONS

We cannot leave the hills without first having a glimpse of the rhododendrons in flower. But for this you will have to visit a hill-station towards the end of March, when it is still winter in the Himalayan foothills.

When, as a boy, I went to a boarding school in Shimla, the little narrow-gauge train would come huffing and puffing up the steep incline towards Summerhill, and we would look out of the windows to admire the scarlet blossoms of the rhododendrons. For eight years of boarding school life we would be brought up the mountains by a little steam-engine, and always the rhododendrons were there to welcome us. Sometimes there was snow on the ground, and the fallen petals would stand out against the snow—scarlet against white.

Dalhousie is especially blessed with rhododendrons; at least it was, when I was last there, fifty years ago! So are the northern slopes of Mussoorie. The hill people call the flower 'baras'; it makes a good wine, if the juice is fermented by a good winemaker.

When Gautam was small, he was having some difficulty with the word 'rhododendron', so to amuse him, I played on the word:

'One rode a horse,' I said, 'and the other rhododendron!'

'Can you ride on a rhododendron?' he asked, not to be fooled.

'Maybe not', I said, 'but you can make jam with it.'

That aroused his interest, and we got his mother to make us rhododendron jam. It went down quite well; but you'll have to ask his mother for the recipe.

As you go further up in the Himalayas, you will find other varieties of the rhododendron—smaller trees with flowers coming in white, yellow, purple, and other shades. These sturdy trees, sometimes forming small forests, can be found up to 8,000 ft in Himachal, Kumaon, Garhwal and Kashmir. Others are found at elevations of 10,000 to 14,000 ft in the Sutlej valley, while the rare white *Rhododendron falconeri* makes its home at a similar altitude in Kashmir.

Now grown as an ornamental tree or shrub in Europe, the rhododendrons are truly at home in the Himalayas.

But why such a long name, Gautam wants to know.

Well it came from the Greek (via Latin) into English: *rhodon* meaning rose + *dendron*, meaning tree. So let's simply call it the Rose Tree!

DEATH OF THE TREES

The peace and quiet of the Maplewood hillside disappeared forever one winter. The powers-that-be decided to build another new road into the mountains and the PWD saw fit to take it right past the cottage, about six feet from the window that overlooked the forest.

In my journal I wrote—Already they have felled most of the trees. The walnut was one of the first to go. A tree I had lived with for over ten years, watching it grow as I had watched Prem's small son Rakesh grow up, looking forward to its new leaf-buds, the broad green leaves or summer turning to spears of gold in September when the walnuts were ripe and ready to fall. I knew this tree better than the others. It was just below the window where a buttress for the road was going up.

Another tree I will miss is the young deodar, the only one growing in this stretch of the woods. Some years back it was stunted from lack of sunlight. The oaks covered it with their shaggy branches, so I cut away some of the overhanging ones and after that the deodar grew much faster. It was just coming into its own this year—now cut down in its prime like my young brother on the road to Delhi last month. Both victims of the roads—the tree kilted by the PWD, my brother by a truck.

Twenty oaks have been felled just in this small stretch near the cottage. By the time this bypass reaches Jabai *khet*, about six miles from here, over a thousand oaks will have been slaughtered, besides many other fine trees—maples, deodars and pines—most of them unnecessarily as they grew some fifty or sixty yards from the roadside.

The trouble is, hardly anyone (with the exception of the contractor who buys the felled trees) really believes that trees and shrubs are necessary. They get in the way so much, don't they? According to my milkman, the only useful tree is the one that can be picked clean of its leaves for fodder! And a young man remarked to me, 'You should come to Pauri. The view is terrific, there's not a tree in the way!'

Well he can stay here now and enjoy the view of the ravaged hillside. But as the oaks have gone, the milkman will have to look further afield for his fodder.

Rakesh calls the maples butterfly trees because when the winged seeds fall, they flutter like butterflies in the breeze. No maples now. No bright red leaves to flame against the sky. No birds! That is to say, no birds near the house. No longer will it be possible for me to open the window and watch the scarlet minivets flitting through the dark green foliage of the oaks... the long-tailed magpies gliding through the trees, the barbet calling insistently from his perch on the top of the deodar. Forest birds, all of them, they will now be in search of some other stretch of surviving forest. The only visitors will be the crows who have learnt to live with and off humans and seem to multiply along with roads, houses and people. And even when all the people have gone, the crows will still be there.

Other things to look forward to—trucks thundering past in the night, perhaps a tea and pakora shop around the corner. The grinding of gears, the music of motor horns. Will the whistling thrush be heard above them? The explosions that continually shatter the silence of the mountains as thousand-year-old rocks are dynamited have frightened away all but the most intrepid of birds and animals. Even the bold langurs haven't shown their faces for over a fortnight.

Somehow, I don't think we shall wait for the tea shop to arrive. There must be some other quiet corner, possibly on the next mountain where new roads have yet to come into being. No doubt this is a negative attitude and if I had any sense I'd open my own tea shop. To retreat is to be a loser. But the trees are losers too and when they fall, they do so with a certain dignity.

Never mind. Men come and go, the mountains remain.

A GOOD PLACE FOR TREES

As my father had told me, Dehra was a good place for trees, and Grandmother's house was surrounded by several kinds—peepul, neem, mango, jackfruit, papaya and an ancient banyan tree. Some of the trees had been planted by my father and grandfather.

'How old is the jackfruit tree?' I asked Grandmother.

'Now let me see,' said Grandmother, looking very thoughtful. 'I should remember the jackfruit tree. Oh yes, your grandfather put it down in 1927. It was during the rainy season. I remember, because it was your father's birthday, and we celebrated it by planting a tree. 14 July 1927. Long before you were born!'

The banyan tree grew behind the house. Its spreading aerial roots which descended into the ground formed a number of twisting passageways in which I liked to wander. The tree was older than the house, older than my grandparents, as old as Dehra. I could hide myself among the roots, behind thick green leaves, and spy on the world below.

It was an enormous tree, about sixty feet high, and the first time I saw it I trembled with excitement because I had never seen such a marvellous tree before. I approached it slowly, even cautiously, as I wasn't sure the tree wanted my friendship. It looked as though it had many secrets. There were sounds and movement in the branches, but I couldn't see who or what made the sounds.

The tree made the first move, the first sign of friendship. It allowed a leaf to fall.

The leaf brushed against my face as it floated down but before it could reach the ground I caught and held it. I studied the leaf, running my fingers over its smooth, glossy surface. Then I put out my hand and touched the rough bark of the tree and this felt good to me. So I removed my shoes and socks, as people do when they enter a holy place, and finding first a foothold and then a handhold on that broad trunk I pulled myself up with the help of the tree's aerial roots.

As I climbed, it seemed as though someone was helping me, that invisible hands, the hands of the spirit in the tree, touched me and helped me climb.

But although the tree wanted me, there were others who were disturbed and alarmed by my arrival. A pair of parrots suddenly shot out of a hole in the trunk and, with shrill cries, flew across the garden, flashes of green and red and gold. A squirrel looked out from behind a branch, saw me, and went scurrying away to inform his friends and relatives.

I climbed higher, looked up, and saw a red beak poised above my head. I shrank away, but the hornbill made no attempt to attack me. He was relaxing in his home, which was a great hole in the tree trunk. Only the bird's head and massive beak were showing. He looked at me in a rather bored way, sleepily opening and shutting his eyes.

'So many creatures live here,' I said to myself. 'I hope none of them is dangerous!' At that moment the hornbill sprang at a passing cricket. Bill and tree trunk met with a loud and resonant 'tonk'!

I was so startled that I nearly fell off the tree. But it was a difficult tree to fall off! It was full of places where one could sit or even lie down. So I moved away from the hornbill, crawled along a branch and so moved quite a distance from the main body of the tree. I left its cold, dark depths for an area penetrated by shafts of sunlight.

No one could see me. I lay flat on the broad branch hidden

by a screen of leaves. People passed by on the road below. A sahib in a sun helmet. His memsahib twirling a coloured sun umbrella. Obviously she did not want to get too brown and be mistaken for a country-born person. Behind them, a pram wheeled along by a nanny.

Then there were a number of locals, some in white dhotis, some in western clothes, some in loincloths. Some with baskets on their heads. Others with coolies to carry their baskets for them. A cloud of dust, the blare of a horn, and down the road, like an out-of-condition dragon, came the latest Morris touring car; then cyclists. Then a man with a basket of papaya balanced on his head. Following him came a man with a performing monkey. This man rattled a little hand drum, and children followed man and monkey along the road. They stopped in the shade of a mango tree on the other side of the road. The little red monkey wore a frilled dress and a baby's bonnet. It danced for the children, while the man sang and played his drum.

The clip-clop of a tonga pony, and Bansi's tonga came rattling down the road. I called down to him, and he reined in with a shout of surprise, and looked up into the branches of the banyan tree.

'What are you doing up there?' he cried.

'Hiding from Grandmother,' I said.

'And when are you coming for that ride?'

'On Tuesday afternoon,' I said.

'Why not today?'

'Ayah won't let me. But she has Tuesdays off.'

Bansi spat red paan juice across the road. 'Your Ayah is jealous,' he said.

'I know,' I said. 'Women are always jealous, aren't they? I suppose it's because she doesn't have a tonga.'

'It's because she doesn't have a tonga driver,' said Bansi, grinning up at me. 'Never mind, I'll come on Tuesday—that's the day after tomorrow, isn't it?'

I nodded down to him, and then started backing along my branch, because I could hear Ayah calling in the distance. Bansi leant forward and smacked his pony across the rump, and the tonga shot forward.

'What are you doing up there?' asked Ayah a little later.

'I was watching a snake cross the road,' I said. I knew she couldn't resist talking about snakes. There weren't as many in Dehra as there had been in Kathiawar, and she was thrilled that I had seen one.

'Was it moving towards you or away from you?' she asked.

'It was going away.'

Ayah's face clouded over. 'That means poverty for the beholder,' she said gloomily.

Later, while scrubbing me down in the bathroom, she began to talk about her dislikes, which included drunkards ('they die quickly, anyway'), misers ('they get murdered sooner or later') and tonga drivers ('they have all the vices').

'You are a very lucky boy,' she said suddenly, peering closely at my tummy.

'Why?' I asked. 'You just said I would be poor because I saw a snake going the wrong way.'

'Well, you won't be poor for long. You have a mole on your tummy, and that's very lucky. And there is one under your armpit, which means you will be famous. Do you have one on the neck? No, thank God! A mole on the neck is the sign of a murderer!'

'Do you have any moles?' I asked.

Ayah nodded seriously and, pulling her sleeve up to her shoulder, showed me a large mole on her arm.

'What does that mean?' I asked.

'It means a life of great sadness,' said Ayah gloomily.

'Can I touch it?' I asked.

'Yes, touch it,' she said, and taking my hand she placed it against the mole.

'It's a nice mole,' I said, wanting to make Ayah happy. 'Can I kiss it?'

'You can kiss it,' said Ayah.

I kissed her on the mole.

'That's nice,' she said.

GENTLE SHADE BY DAY

Those who have spent time in non-air-conditioned parts of India will remember with gratitude those gracious trees that provide shade and shelter during the summer months—the banyan, peepul, mango, neem and others. Coastal dwellers are not so fortunate for there is not much shade to be had from a palm tree unless you keep moving in its long but insubstantial shadow.

I am not surprised that the sages of old were given to sitting beneath the peepul tree. They might have had various religious reasons for calling it sacred but I am sure there was a good practical reason as well. Few trees provide a cooler shade than it does. Even on the stillest of days, the peepul leaves are forever twirling and with thousands of leaves spinning like tops, there is quite a breeze for anyone sitting below.

However, there are warnings about peepul trees—'Gentle shade by day and terror by night!' During the night, the tree is said to be alive with various spirits, most of them inimical to man. One is advised not to sleep beneath it, for this is construed by a ghost as an invitation to jump down your throat and take possession of you, or at the very least, ruin your digestion.

It is also said to be unlucky to sleep beneath a tamarind, but I have often reclined in the pleasant shade of this noble tree and have come to no harm. A famous tamarind stands over the tomb of Tansen, the great musician and singer of Akbar's court at Gwalior. Its leaves, though bitter, are eaten by singers to improve their voices.

A mango grove is a wonderful place for an afternoon siesta. But if the mangoes are ripening, there is usually a great deal of activity going on with parrots, crows, monkeys and small boys, all attempting to evade the watchman who uses an empty kerosene tin as a drum to try and frighten them away. So it's not the ideal place for a nap then, but the shade under a mango grove is dark, deep and very soothing.

The banyan tree with its aerial roots represents the matted hair of Lord Shiva. There is always shade and space beneath a venerable old banyan. It is still a popular community centre in our Indian villages but is becoming a rarity in cities simply because it covers so large an area. And if you cut its aerial roots, the tree topples over. Other handsome trees related to the banyan are the pilkhan and the chilkhan, large spreading evergreens, both quite noticeable on some of New Delhi's wider avenues.

The neem is a tall tree, but its numerous branches give it a shady head. One of my greatest pleasures is to walk beneath an avenue of neem trees after a shower of rain. As the fallen berries are crushed underfoot, they give out a sharp heady fragrance, which I find exhilarating. Apart from its medicinal uses, the tree is connected in legends with the Sun God, as in the story of *Neembarak*, 'The Sun in a Neem Tree', who invited to dinner a Bairagi tribal whose rules forbade him to eat anything except by daylight. When dinner was delayed after sundown, Suraj Narayan, the Sun God, obligingly descended from a neem tree and continued shining till dinner was over.

On this pleasant note I end this tribute, only adding that shade-giving trees symbolize the harmony between man and nature and that our ancestors, in their devotion to trees and reverence for them, clearly showed that they knew what was good for them.

THE CHERRY TREE

One day, when Rakesh was six, he walked home from the Mussoorie bazaar eating cherries. They were a little sweet, a little sour; small, bright red cherries that had come all the way from the Kashmir Valley.

Here in the Himalayan foothills where Rakesh lived, there were not many fruit trees. The soil was stony, and the dry cold winds stunted the growth of most plants. But on the more sheltered slopes there were forests of oak and deodar.

Rakesh lived with his grandfather on the outskirts of Mussoorie, just where the forest began. His father and mother lived in a small village fifty miles away, where they grew maize and rice and barley in narrow terraced fields on the lower slopes of the mountain. But there were no schools in the village, and Rakesh's parents were keen that he should go to school. As soon as he was of school-going age, they sent him to stay with his grandfather in Mussoorie.

Grandfather was a retired forest ranger. He had a little cottage outside the town.

Rakesh was on his way home from school when he bought the cherries. He paid fifty paise for the bunch. It took him about half an hour to walk home, and by the time he reached the cottage there were only three cherries left.

'Have a cherry, Grandfather,' he said, as soon as he saw his grandfather in the garden.

Grandfather took one cherry and Rakesh promptly ate the other two. He kept the last seed in this mouth for some time, rolling it round and round on his tongue until all the tang had gone. Then he placed the seed on the palm of his hand and studied it.

'Are cherry seeds lucky?' asked Rakesh.

'Of course.'

'Then I'll keep it.'

'Nothing is lucky if you put it away. If you want luck, you must put it to some use.'

'What can I do with a seed?'

'Plant it.'

So Rakesh found a small shade and began to dig up a flower bed.

'Hey, not there,' said Grandfather. 'I've sown mustard in that bed. Plant it in that shady corner where it won't be disturbed.'

Rakesh went to a corner of the garden where the earth was soft and yielding. He did not have to dig. He pressed the seed into the soil with his thumb and it went right in.

Then he had his lunch and ran off to play cricket with his friends and forgot all about the cherry seed.

When it was winter in the hills, a cold wind blew down from the snows and went *whoo-whoo-whoo* through the deodar trees, and the garden was dry and bare. In the evenings, Grandfather told Rakesh stories—stories about people who turned into animals, and ghosts who lived in trees, and beans that jumped and stones that wept—and in turn Rakesh would read to him from the newspaper, Grandfather's eyesight being rather weak. Rakesh found the newspaper very dull—especially after the stories—but Grandfather wanted all the news…

They knew it was spring when the wild duck flew north again, to Siberia. Early in the morning, when he got up to chop wood and light a fire, Rakesh saw the V-shaped formation streaming northwards, the calls of the birds carrying clearly through the thin mountain air.

One morning in the garden, he bent to pick up what he thought was a small twig and found to his surprise that it was well-rooted. He stared at it for a moment, then ran to fetch Grandfather, calling, 'Dada, come and look, the cherry tree has come up!'

'What cherry tree?' asked Grandfather, who had forgotten about it.

'The seed we planted last year—look, it's come up!'

Rakesh went down on his haunches, while Grandfather bent almost double and peered down at the tiny tree. It was about four inches high.

'Yes, it's a cherry tree,' said Grandfather. 'You should water it now and then.'

Rakesh ran indoors and came back with a bucket of water.

'Don't drown it!' said Grandfather.

Rakesh gave it a sprinkling and circled it with pebbles.

'What are the pebbles for?' asked Grandfather.

'For privacy,' said Rakesh.

He looked at the tree every morning but it did not seem to be growing very fast. So, he stopped looking at it—except quickly, out of the corner of his eye. And, after a week or two, when he allowed himself to look at it properly, he found that it had grown—at least an inch!

That year the monsoon rains came early and Rakesh plodded to and from school in raincoat and gum boots. Ferns sprang from the trunks of trees, strange-looking lilies came up in the long grass, and even when it wasn't raining the trees dripped, and mist came curling up the valley. The cherry tree grew quickly in this season.

It was about two feet high when a goat entered the garden and ate all the leaves. Only the main stem and two thin branches remained.

'Never mind,' said Grandfather, seeing that Rakesh was upset. 'It will grow again, cherry trees are tough.'

Towards the end of the rainy season new leaves appeared on the tree. Then a woman cutting grass scrambled down the hillside, her scythe swishing through the heavy monsoon foliage. She did not try to avoid the tree: one sweep, and the cherry tree was cut in two.

When Grandfather saw what had happened, he went after the woman and scolded her; but the damage could not be repaired.

'Maybe it will die now,' said Rakesh.

'Maybe,' said Grandfather.

But the cherry tree had no intention of dying.

By the time summer came round again, it had sent out several new shoots with tender green leaves. Rakesh had grown taller too. He was eight now, a sturdy boy with curly black hair and deep black eyes. Blackberry eyes, Grandfather called them.

That monsoon Rakesh went home to his village, to help his father and mother with the planting and ploughing and sowing. He was thinner but stronger when he came back to Grandfather's house at the end of the rains, to find that the cherry tree had grown another foot. It was now up to his chest.

Even when there was rain, Rakesh would sometimes water the tree. He wanted it to know that he was there.

One day he found a bright green praying mantis perched on a branch, peering at him with bulging eyes. Rakesh let it remain there. It was the cherry tree's first visitor.

The next visitor was a hairy caterpillar, who started making a meal of the leaves. Rakesh removed it quickly and dropped it on a heap of dry leaves.

'They're pretty leaves,' said Rakesh. 'And they are always ready to dance. If there's a breeze.'

After Grandfather had come indoors, Rakesh went into the garden and lay down on the grass beneath the tree. He gazed up through the leaves at the great blue sky; and turning on his side, he could see the mountain striding away into the clouds. He was still lying beneath the tree when the evening shadows crept across the garden. Grandfather came back and sat down beside Rakesh, and they waited in silence until the stars came out and the nightjar began to call. In the forest below, the crickets and cicadas began tuning up; and suddenly the tree was full of the sound of insects.

'There are so many trees in the forest,' said Rakesh. 'What's so special about this tree? Why do we like it so much?'

'We planted it ourselves,' said Grandfather. 'That's why it's special.'

'Just one small seed,' said Rakesh, and he touched the smooth bark of the tree that had grown. He ran his hand along the trunk of the tree and put his finger to the tip of a leaf. 'I wonder,' he whispered, 'is this what it feels like to be God?'

UNDER THE DEODARS

The names of trees often find their way into the titles of books and stories. Kipling's earliest stories were published in India under the title *Under the Deodars*—for it was under the deodars of Shimla that all the scandals, intrigues, romances and hauntings took place!

As you ascend the Himalayan foothills, the trees of the plains give way to the trees of the mountains, and at 7,000 ft you will be welcomed by the deodars (*Cedrus deodara*), singly or in large numbers, for these great trees are indigenous to the Himalayas. In the presence of deodars, the air is purer, more bracing. They will give you a sense of renewal.

Deodar comes from *devdar*, Sanskrit for 'Tree of God'. And indeed no tree could be nobler, growing to a great height and living for two or three hundred years, sometimes longer. Some of the oldest and most magnificent specimens are found in the premises of temples, where they are protected. These great trees are, at their most luxuriant, around 8,000 ft.

A similar tree is the cedar of Lebanon, which grows on the mountains of Lebanon and Syria. Some are reputed to be about 2,000 years old with a girth of some 40 ft. Legend has it that it provided the timber used by King Solomon for building his temple.

Mr V.P. Mehta tells me that at the Forest Research Institute in Dehradun, a cross section of an old deodar shows 700 distinct annual growth rings. The tree was born some time during the reign of Alauddin Khilji in the twelfth century, and lived through the Mughal period and the great part of the British rule in India.

If only trees could talk, what great historians they would be!

THE FRIENDLY OAKS

When I remember the oak trees below the old cottage in the hills, I see again the long-tailed blue magpies fluttering from tree to tree; a woodpecker tapping away at the bark; brightly coloured minivets standing out against its dark foliage; and a band of langurs crashing through the branches in search of acorns. The oaks were always hospitable to birds, beasts and insects.

This is the Banj oak, growing from 5,000 ft to 7,000 ft in the western Himalayas. Higher up, other species of oak take over, dominating the most northern slopes facing the snows.

Oaks like each other's company, but each one has a certain individuality—its branches growing as they please, avoiding symmetry, taking on shapes that often give the tree an untidy look, as though it could do with a little discipline. But that of course, is its charm.

Our Himalayan oaks are a little different from the famous oaks of England and Europe. The mighty oak is England's noblest tree. The Romans made the crown of oak leaves (which symbolizes bravery) their principal award, and it could only be given to a citizen who had slain an enemy, won a battle or saved the life of a fellow citizen. The Celts worshipped the oak, regarding it as the symbol of their most prized virtue, hospitality.

Even the moss grows more readily on the trunks of an old oak tree.

GREAT TREES OF GARHWAL

Living for many years in a cottage at 7,000 ft in the Garhwal Himalayas, I was fortunate to have a big window that opened out on the forest, so that the trees were almost within my reach. Had I jumped, I should have landed quite safely in the arms of an oak or chestnut.

The incline of the hill was such that my first floor window opened on what must, I suppose, have been the second floor of the tree. I never made the jump, but the big langurs—silver-grey monkeys with long swishing tails—often leapt from the trees onto the corrugated tin roof and made enough noise to disturb the bats sleeping in the space between the roof and ceiling.

Standing on its own was a walnut tree, and truly this was a tree for all seasons. In winter the branches were bare; but they were smooth and straight and round like the arms of a woman in a painting by Jamini Roy. In the spring, each branch produced a hard, bright spear of new leaf. By midsummer the entire tree was in leaf; and towards the end of the monsoon, the walnuts, encased in their green jackets, had reached maturity.

Then the jackets began to split, revealing the hard brown shell of the walnuts. Inside the shell was the nut itself. Look closely at the nut and you will notice that it is shaped rather like the human brain. No wonder the ancients prescribed walnuts for headaches!

Every year the tree gave me a basket of walnuts. But last year the walnuts were disappearing one by one, and I was at a loss to know who had been taking them. Could it have been Bijju, the milkman's son? He was an inveterate tree climber.

But he was usually to be found on oak trees, gathering fodder for his cows. He told me that his cows liked oak leaves but did not care for walnuts. He admitted that they had relished my dahlias, which they had eaten the previous week, but he denied having fed them walnuts.

It wasn't the woodpecker. He was out there every day, knocking furiously against the bark of the tree, trying to prise an insect out of a narrow crack. He was strictly non-vegetarian and none the worse for it.

One day I found a fat langur sitting in the walnut tree. I watched him for some time to see if he was going to help himself to the nuts, but he was only sunning himself. When he thought I wasn't looking, he came down and ate the geraniums; but he did not take any walnuts.

The walnuts had been disappearing early in the morning while I was still in bed. So one morning I surprised everyone, including myself, by getting up before sunrise. I was just in time to catch the culprit climbing out of the walnut tree.

She was an old woman, who sometimes came to cut grass on the hillside. Her face was as wrinkled as the walnuts she had been helping herself to. In spite of her age, her arms and legs were sturdy. When she saw me, she was as swift as a civet cat in getting out of the tree.

'And how many walnuts did you gather today, Grandmother?' I asked.

'Only two,' she said with a giggle, offering them to me on her open palm. I accepted one of them. Encouraged, she climbed back into the tree and helped herself to the remaining nuts. It was impossible to object. I was taken up in admiration of her agility in the tree. She must have been about sixty, and I was a mere forty-five, but I knew I would never be climbing trees again.

To the victor the spoils!

The horse chestnuts are inedible, even the monkeys throw them away in disgust. Once, on passing beneath a horse chestnut tree, a couple of chestnuts bounched off my head. Looking up, I saw that they had been dropped on me by a couple of mischievous rhesus monkeys.

The tree itself is a friendly one, especially in summer when it is in full leaf. The least breath of wind makes the leaves break into conversation, and their rustle is a cheerful sound, unlike the sad notes of pine trees in the wind. The spring flowers look like candelabra, and when the blossoms fall they carpet the hillside with their pale pink petals.

We pass now to my favourite tree, the deodar. In Garhwal and Kumaon it is called Dujar or Devdar; in Jaunsar and parts of Himachal it is known as the Kelu or Kelon. It is also identified with the cedar of Lebanon (the cones are identical), although the deodar's needles are slightly longer and more bluish. Trees, like humans, change with their environment. Several persons familiar with the deodar at Indian hill stations, when asked to point it out in London's Kew Gardens, indicated the cedar of Lebanon; and when shown a deodar, declared that they had never seen this tree in the Himalayas!

We shall stick to the name deodar, which comes from the Sanskrit *deva-daru* (divine tree). It is a sacred tree in the Himalayas; not worshipped, not protected in the way that a peepul is in the plains, but sacred in that its timber has always been used in temples, for doors, windows, walls and even roofs. Quite frankly, I would just as soon worship the deodar as worship anything, for in its beauty and majesty it represents Nature in its most noble aspect.

No one who has lived amongst deodars would deny that it is the most godlike of Himalayan trees. It stands erect, dignified;

and though in a strong wind it may hum and sigh and moan, it does not bend to the wind. The snow slips softly from its resilient branches. In the spring the new leaves are tender green, while during the monsoon the tiny young cones spread like blossoms in the dark green folds of the branches. The deodar thrives in the rain and enjoys the company of its own kind. Where one deodar grows, there will be others. Isolate a young tree and it will often pine away.

The great deodar forests are found along the upper reaches of the Bhagirathi valley and the Tons in Garhwal; and in Himachal and Kashmir, along the Chenab and the Jhelum, and also the Kishenganga; it is at its best between 7,000 and 9,000 ft. I had expected to find it on the upper reaches of Alaknanda, but could not find a single deodar along the road to Badrinath. That particular valley seems hostile to trees in general, and deodars in particular.

The average girth of the deodar is 15–20 ft, but individual trees often attain a great size. Records show that one great deodar was 250 ft high, 20 ft in girth at the base, and more than five hundred and fifty years old. The timber of these trees, which is unaffected by extremes of climate, was always highly prized for house buildings; and in the villages of Jaunsar Bawar, finely carved doors and windows are a feature of the timbered dwellings. Many of the quaint old bridges over the Jhelum in Kashmir are supported on pillars fashioned from whole deodar trees; some of these bridges are more than five hundred years old.

To return to my own trees, I went among them often, acknowledging their presence with the touch of my hand against their trunks—the walnut's smooth and polished; the pine's patterned and whorled; and oak's rough, gnarled, full of experience. The oak had been there the longest, and the wind had bent his upper branches and twisted a few, so that he looked shaggy and undistinguished. It is a good tree for the privacy of birds, its crooked branches spreading out with no particular effect; and sometimes the tree seems uninhabited until there is a whirring sound, as of a helicopter approaching, and a party of long-tailed blue magpies stream across the forest glade.

After the monsoon, when the dark red berries had ripened on the hawthorn, this pretty tree was visited by green pigeons, the kokla birds of Garhwal, who clambered upside-down among the fruit-laden twigs. And during winter, a white-capped redstart perched on the bare branches of the wild pear tree and whistled cheerfully. He had come down from higher places to winter in the garden.

The pines grow on the next hill—the chir, the Himalayan blue pine, and the long-leaved pine—but there is a small blue pine a little way below the cottage, and sometimes I sit beneath it to listen to the wind playing softly in its branches.

Open the window at night and there is usually something to listen to: the mellow whistle of a pigmy owlet, or the cry of a barking deer that has scented the proximity of a panther. Sometimes, if you are lucky, you will see the moon coming up over Nag Tibba and two distant deodars in perfect silhouette.

Some sounds cannot be recognized. They are strange night sounds, the sounds of the trees themselves, stretching their limbs in the dark, shifting a little, flexing their fingers. Great trees of the mountains, they know me well. They know my face in the window; they see me watching them, watching them grow, listening to their secrets, bowing my head before their outstretched arms and seeking their benediction.

THE PROSPECT OF FLOWERS

Fern Hill, The Oaks, Hunter's Lodge, The Parsonage, The Pines, Dumbarnie, Mackinnon's Hall and Windermere. These are the names of some of the old houses that still stand on the outskirts of one of the smaller Indian hill stations. Most of them have fallen into decay and ruin. They are very old, of course—built over a hundred years ago by Britons who sought relief from the searing heat of the plains. Today's visitors to the hill stations prefer to live near the markets and cinemas and many of the old houses, set amid oak and maple and deodar, are inhabited by wild cats, bandicoots, owls, goats and the occasional charcoal burner or mule driver.

But among these neglected mansions stands a neat, whitewashed cottage called Mulberry Lodge. And in it, up to a short time ago, lived an elderly English spinster named Miss Mackenzie.

In years Miss Mackenzie was more than 'elderly', being well over eighty. But no one would have guessed it. She was clean, sprightly and wore old-fashioned but well-preserved dresses. Once a week, she walked the two miles to town to buy butter and jam and soap, and sometimes a small bottle of eau de cologne.

She had lived in the hill station since she had been a girl in her teens, and that had been before the First World War. Though she had never married, she had experienced a few love affairs and was far from being the typical frustrated spinster of fiction. Her parents had been dead thirty years; her brother and sister were also dead. She had no relatives in India and she lived on a small pension of forty rupees a month and the gift parcels that were sent out to her from New Zealand by a friend of her youth.

Like other lonely old people, she kept a pet—a large black cat with bright yellow eyes. In her small garden she grew dahlias, chrysanthemums, gladioli and a few rare orchids. She knew a great deal about plants and about wild flowers, trees, birds and insects. She had never made a serious study of these things, but having lived with them for so many years had developed an intimacy with all that grew and flourished around her.

She had few visitors. Occasionally, the padre from the local church called on her, and once a month the postman came with a letter from New Zealand or her pension papers. The milkman called every second day with a litre of milk for the lady and her cat. And sometimes she received a couple of eggs free, for the egg seller remembered a time when Miss Mackenzie, in her earlier prosperity, had bought eggs from him in large quantities. He was a sentimental man. He remembered her as a ravishing beauty in her twenties when he had gazed at her in round-eyed, nine-year-old wonder and consternation.

Now it was September and the rains were nearly over and Miss Mackenzie's chrysanthemums were coming into their own. She hoped the coming winter wouldn't be too severe because she found it increasingly difficult to bear the cold.

One day, as she was pottering about in her garden, she saw a schoolboy plucking wild flowers on the slope above the cottage.

'Who's that?' she called. 'What are you up to, young man?'

The boy was alarmed and tried to dash up the hillside, but he slipped on pine needles and came slithering down the slope on to Miss Mackenzie's nasturtium bed.

When he found there was no escape, he gave a bright disarming smile and said, 'Good morning, miss.'

He belonged to the local English-medium school and wore a bright red blazer, and a red and black striped tie. Like most polite Indian schoolboys, he called every woman 'miss'.

'Good morning,' said Miss Mackenzie severely. 'Would you mind moving out of my flower bed?'

The boy stepped gingerly over the nasturtiums and looked up at Miss Mackenzie with dimpled cheeks and appealing eyes. It was impossible to be angry with him.

'You're trespassing,' said Miss Mackenzie.

'Yes, miss.'

'And you ought to be in school at this hour.'

'Yes, miss.'

'Then what are you doing here?'

'Picking flowers, miss.' And he held up a bunch of ferns and wild flowers.

'Oh,' Miss Mackenzie was disarmed. It was a long time since she had seen a boy taking an interest in flowers and, what was more, playing truant from school in order to gather them.

'Do you like flowers?' she asked.

'Yes, miss. I'm going to be a botan—a botantist?'

'You mean a botanist.'

'Yes, miss.'

'Well, that's unusual. Most boys at your age want to be pilots or soldiers or perhaps engineers. But you want to be a botanist. Well, well. There's still hope for the world, I see. And do you know the names of these flowers?'

'This is a *bukhilo* flower,' he said, showing her a small golden flower. 'That's a Pahari name. It means puja or prayer. The flower is offered during prayers. But I don't know what this is…'

He held out a pale pink flower with a soft, heart-shaped leaf.

'It's a wild begonia,' said Miss Mackenzie. 'And that purple stuff is salvia, but it isn't wild. It's a plant that escaped from my garden. Don't you have any books on flowers?'

'No, miss.'

'All right, come in and I'll show you a book.'

She led the boy into a small front room, which was crowded with furniture and books and vases and jam jars, and offered him a chair. He sat awkwardly on its edge. The black cat immediately leapt on to his knees and settled down on them, purring loudly.

'What's your name?' asked Miss Mackenzie, as she rummaged through her books.

'Anil, miss.'

'And where do you live?'

'When school closes, I go to Delhi. My father has a business.'

'Oh, and what's that?'

'Bulbs, miss.'

'Flower bulbs?'

'No, electric bulbs.'

'Electric bulbs! You might send me a few, when you get home. Mine are always fusing and they're so expensive, like everything else these days. Ah, here we are!' She pulled a heavy volume down from the shelf and laid it on the table. '*Flora Himaliensis*, published in 1892, and probably the only copy in India. This is a very valuable book, Anil. No other naturalist has recorded so many wild Himalayan flowers. And let me tell you this, there are many flowers and plants which are still unknown to the fancy botanists who spend all their time with microscopes instead of in the mountains. But perhaps, *you'll* do something about that, one day.'

'Yes, miss.'

They went through the book together, and Miss Mackenzie pointed out many flowers that grew in and around the hill station while the boy made notes of their names and seasons. She lit a stove and put the kettle on for tea. And then the old English lady and the small Indian boy sat side by side over cups of hot sweet tea, absorbed in a book on wild flowers.

'May I come again?' asked Anil, when finally he rose to go.

'If you like,' said Miss Mackenzie. 'But not during school hours. You mustn't miss your classes.'

After that, Anil visited Miss Mackenzie about once a week, and nearly always brought a wild flower for her to identify. She found herself looking forward to the boy's visits—and sometimes, when more than a week passed and he didn't come, she was disappointed and lonely and would grumble at the black cat.

Anil reminded her of her brother, when the latter had been a boy. There was no physical resemblance. Andrew had been fair-haired and blue-eyed. But it was Anil's eagerness, his alert, bright look and the way he stood—legs apart, hands on hips, a picture of confidence—that reminded her of the boy who had shared her own youth in these same hills.

And why did Anil come to see her so often? Partly because she knew about wild flowers and he really did want to become a botanist. And partly because she smelt of freshly baked bread and that was a smell his own grandmother had possessed. And partly because she was lonely and sometimes a boy of twelve can sense loneliness better than an adult. And partly because he was a little different from other children.

By the middle of October, when there was only a fortnight left for the school to close, the first snow had fallen on the distant mountains. One peak stood high above the rest, a white pinnacle against the azure blue sky. When the sun set, this peak turned from orange to gold to pink to red.

'How high is that mountain?' asked Anil.

'It must be over twelve thousand feet,' said Miss Mackenzie. 'About thirty miles from here, as the crow flies. I always wanted to go there, but there was no proper road. At that height, there'll be flowers that you don't get here—the blue gentian and the purple columbine, the anemone and the edelweiss.'

'I'll go there one day,' said Anil.

'I'm sure you will, if you really want to.'

The day before his school closed, Anil came to say goodbye to Miss Mackenzie.

'I don't suppose you'll be able to find many wild flowers in Delhi,' she said. 'But have a good holiday.'

'Thank you, miss.'

As he was about to leave, Miss Mackenzie, on an impulse, thrust the *Flora Himaliensis* into his hands.

'You keep it,' she said. 'It's a present for you.'

'But I'll be back next year, and I'll be able to look at it then. It's so valuable.'

'I know it's valuable and that's why I've given it to you. Otherwise it will only fall into the hands of the junk dealers.'

'But, miss...'

'Don't argue. Besides, I may not be here next year.'

'Are you going away?'

'I'm not sure. I may go to England.'

She had no intention of going to England; she had not seen the country since she was a child, and she knew she would not fit in with the life of post-war Britain. Her home was in these hills, among the oaks and maples and deodars. It was lonely, but at her age it would be lonely anywhere.

The boy tucked the book under his arm, straightened his tie, stood stiffly to attention and said, 'Goodbye, Miss Mackenzie.' It was the first time he had spoken her name.

Winter set in early and strong winds brought rain and sleet, and soon there were no flowers in the garden or on the hillside. The cat stayed indoors, curled up at the foot of Miss Mackenzie's bed. Miss Mackenzie wrapped herself up in all her old shawls and mufflers but still she felt the cold. Her fingers grew so stiff that she took almost an hour to open a can of baked beans. And then it snowed and for several days the milkman did not come. The postman arrived with her pension papers but she felt too tired to take them up to town to the bank.

She spent most of the time in bed. It was the warmest place. She kept a hot-water bottle at her back and the cat kept her feet warm. She lay in bed, dreaming of the spring and summer months. In three months' time the primroses would be out and with the coming of spring the boy would return.

One night the hot-water bottle burst and the bedding was soaked through. As there was no sun for several days, the blanket remained damp. Miss Mackenzie caught a chill and had to keep to her cold, uncomfortable bed. She knew she had a fever but there was no thermometer with which to take her temperature. She had difficulty breathing.

A strong wind sprang up one night and the window flew open and kept banging all night. Miss Mackenzie was too weak to get up and close it, and the wind swept the rain and sleet into the room. The cat crept into the bed and snuggled close to its mistress's warm body. But towards morning that body had lost its warmth and the cat left the bed and started scratching about on the floor.

As a shaft of sunlight streamed through the open window, the milkman arrived. He poured some milk into the cat's saucer on the doorstep and the cat leapt down from the windowsill and made for the milk.

The milkman called a greeting to Miss Mackenzie, but received no answer. Her window was open and he had always known her to be up before sunrise. So he put his head in at the window and called again. But Miss Mackenzie did not answer. She had gone away to the mountain where the blue gentian and purple columbine grew.

THE FRIENDLY BANYAN

It's the hour of cow dust.
A slanting sunbeam strikes
Through the gathering mist
And turns the dust to gold.
The grazing cattle stream home.
The wading egrets seek shelter,
And in the over-arching banyan tree
The mynas squabble, the squirrels play
The fruit-bats come to life;
And then the sun sinks in the west,
And in the friendly banyan tree there's rest.

I wrote these lines only a few minutes ago. I had only to think of the majestic banyan tree, and I found myself breaking into verse!

As a boy, how I loved to explore the passageways created by the tree's aerial roots—those spreading branches that hung to the ground and created fresh roots. The banyan tree will keep spreading, if you will allow it to do so. But they do need plenty of space—the outskirts of a village, the banks of a pond, the centre of a park. Don't disturb those aerial roots. If you cut them away, this mighty tree might well topple over. Those roots are like the pillars that support a temple.

ANGRY RIVER

In the middle of the big river, the river that began in the mountains and ended in the sea, was a small island. The river swept round the island, sometimes clawing at its banks, but never going right over it. It was over twenty years since the river had flooded the island, and at that time no one had lived there. But for the last ten years a small hut had stood there, a mud-walled hut with a sloping thatched roof. The hut had been built into a huge rock, so only three of the walls were mud, and the fourth was rock.

Goats grazed on the short grass which grew on the island, and on the prickly leaves of thorn bushes. A few hens followed them about. There was a melon patch and a vegetable patch. In the middle of the island stood a peepul tree. It was the only tree there. Even during the Great Flood, when the island had been under water, the tree had stood firm.

It was an old tree. A seed had been carried to the island by a strong wind some fifty years back, had found shelter between two rocks, had taken root there, and had sprung up to give shade and shelter to a small family; and Indians love peepul trees, especially during the hot summer months when the heart-shaped leaves catch the least breath of air and flutter eagerly, fanning those who sit beneath.

A sacred tree, the peepul: the abode of spirits, good and bad.

'Don't yawn when you are sitting beneath the tree,' Grandmother used to warn Sita.

'And if you must yawn, always snap your fingers in front of your mouth. If you forget to do that, a spirit might jump down your throat!'

'And then what will happen?' asked Sita.

'It will probably ruin your digestion,' said Grandfather, who wasn't much of a believer in spirits.

The peepul had a beautiful leaf, and Grandmother likened it to the body of the mighty god Krishna—broad at the shoulders, then tapering down to a very slim waist.

It was an old tree, and an old man sat beneath it. He was mending a fishing net. He had fished in the river for ten years, and he was a good fisherman. He knew where to find the slim silver chilwa fish and the big, beautiful mahseer and the long-moustached singhara; he knew where the river was deep and where it was shallow; he knew which baits to use—which fish liked worms and which liked gram. He had taught his son to fish, but his son had gone to work in a factory in a city, nearly a hundred miles away. He had no grandson; but he had a granddaughter, Sita, and she could do all the things a boy could do, and sometimes she could do them better. She had lost her mother when she was very small. Grandmother had taught her all the things a girl should know, and she could do these as well as most girls. But neither of her grandparents could read or write, and as a result Sita couldn't read or write either.

There was a school in one of the villages across the river, but Sita had never seen it. There was too much to do on the island.

While Grandfather mended his net, Sita was inside the hut, pressing her grandmother's forehead, which was hot with fever. Grandmother had been ill for three days and could not eat. She had been ill before, but she had never been so bad. Grandfather had brought her some sweet oranges from the market in the nearest town, and she could suck the juice from the oranges, but she couldn't eat anything else.

She was younger than Grandfather, but because she was

sick, she looked much older. She had never been very strong.

When Sita noticed that Grandmother had fallen asleep, she tiptoed out of the room on her bare feet and stood outside.

The sky was dark with monsoon clouds. It had rained all night, and in a few hours it would rain again. The monsoon rains had come early, at the end of June. Now it was the middle of July, and already the river was swollen. Its rushing sound seemed nearer and more menacing than usual.

Sita went to her grandfather and sat down beside him beneath the peepul tree.

'When you are hungry, tell me,' she said, 'and I will make the bread.'

'Is your grandmother asleep?'

'She sleeps. But she will wake soon, for she has a deep pain.'

The old man stared out across the river, at the dark green of the forest, at the grey sky, and said, 'Tomorrow, if she is not better, I will take her to the hospital at Shahganj. There they will know how to make her well. You may be on your own for a few days—but you have been on your own before...'

Sita nodded gravely; she had been alone before, even during the rainy season. Now she wanted Grandmother to get well, and she knew that only Grandfather had the skill to take the small dugout boat across the river when the current was so strong. Someone would have to stay behind to look after their few possessions.

Sita was not afraid of being alone, but she did not like the look of the river. That morning, when she had gone down to fetch water, she had noticed that the level had risen. Those rocks which were normally spattered with the droppings of snipe and curlew and other water birds had suddenly disappeared.

They disappeared every year—but not so soon, surely?

'Grandfather, if the river rises, what will I do?'

'You will keep to the high ground.'

'And if the water reaches the high ground?'
'Then take the hens into the hut, and stay there.'
'And if the water comes into the hut?'
'Then climb into the peepul tree. It is a strong tree. It will not fall. And the water cannot rise higher than the tree!'
'And the goats, Grandfather?'
'I will be taking them with me, Sita. I may have to sell them to pay for good food and medicines for your grandmother. As for the hens, if it becomes necessary, put them on the roof. But do not worry too much'—and he patted Sita's head—'the water will not rise as high. I will be back soon, remember that.'
'And won't Grandmother come back?'
'Yes, of course, but they may keep her in the hospital for some time.'

◆

Towards evening, it began to rain again—big pellets of rain, scarring the surface of the river. But it was warm rain, and Sita could move about in it. She was not afraid of getting wet; she rather liked it. In the previous month, when the first monsoon shower had arrived, washing the dusty leaves of the tree and bringing up the good smell of the earth, she had exulted in it, had run about shouting for joy. She was used to it now, and indeed a little tired of the rain, but she did not mind getting wet. It was steamy indoors, and her thin dress would soon dry in the heat from the kitchen fire.

She walked about barefooted, bare-legged. She was very sure on her feet; her toes had grown accustomed to gripping all kinds of rocks, slippery or sharp. And though thin, she was surprisingly strong.

Black hair streaming across her face. Black eyes. Slim brown arms. A scar on her thigh—when she was small, visiting her mother's village, a hyena had entered the house where she was sleeping, fastened on to her leg and tried to drag her away, but her screams had roused the villagers and the hyena had run off.

She moved about in the pouring rain, chasing the hens into a shelter behind the hut. A harmless brown snake, flooded out of its hole, was moving across the open ground. Sita picked up a stick, scooped the snake up, and dropped it between a cluster of rocks. She had no quarrel with snakes. They kept down the rats and the frogs. She wondered how the rats had first come to the island—probably in someone's boat, or in a sack of grain. Now it was a job to keep their numbers down.

When Sita finally went indoors, she was hungry. She ate some dried peas and warmed up some goat's milk. Grandmother woke once and asked for water, and Grandfather held the brass tumbler to her lips.

◆

It rained all night.

The roof was leaking, and a small puddle formed on the floor. They kept the kerosene lamp alight. They did not need the light, but somehow it made them feel safer.

The sound of the river had always been with them, although they were seldom aware of it; but that night they noticed a change in its sound. There was something like a moan, like a wind in the tops of tall trees and a swift hiss as the water swept round the rocks and carried away pebbles. And sometimes there was a rumble, as loose earth fell into the water.

Sita could not sleep.

She had a rag doll, made with Grandmother's help out of bits of old clothing. She kept it by her side every night. The doll was someone to talk to, when the nights were long and sleep elusive. Her grandparents were often ready to talk—and

Grandmother, when she was well, was a good storyteller—but sometimes Sita wanted to have secrets, and though there were no special secrets in her life, she made up a few, because it was fun to have them. And if you have secrets, you must have a friend to share them with, a companion of one's own age. Since there were no other children on the island, Sita shared her secrets with the rag doll whose name was Mumta.

Grandfather and Grandmother were asleep, though the sound of Grandmother's laboured breathing was almost as persistent as the sound of the river.

'Mumta,' whispered Sita in the dark, starting one of her private conversations.

'Do you think Grandmother will get well again?'

Mumta always answered Sita's questions, even though the answers could only be heard by Sita.

'She is very old,' said Mumta.

'Do you think the river will reach the hut?' asked Sita.

'If it keeps raining like this, and the river keeps rising, it will reach the hut.'

'I am a little afraid of the river, Mumta. Aren't you afraid?'

'Don't be afraid. The river has always been good to us.'

'What will we do if it comes into the hut?'

'We will climb on to the roof.'

'And if it reaches the roof?'

'We will climb the peepul tree. The river has never gone higher than the peepul tree.'

As soon as the first light showed through the little skylight, Sita got up and went outside. It wasn't raining hard, it was drizzling, but it was the sort of drizzle that could continue for days, and it probably meant that heavy rain was falling in the hills where the river originated.

Sita went down to the water's edge. She couldn't find her

favourite rock, the one on which she often sat dangling her feet in the water, watching the little chilwa fish swim by. It was still there, no doubt, but the river had gone over it.

She stood on the sand, and she could feel the water oozing and bubbling beneath her feet.

The river was no longer green and blue and flecked with white, but a muddy colour.

She went back to the hut. Grandfather was up now. He was getting his boat ready.

Sita milked a goat. Perhaps it was the last time she would milk it.

◆

The sun was just coming up when Grandfather pushed off in the boat. Grandmother lay in the prow. She was staring hard at Sita, trying to speak, but the words would not come. She raised her hand in a blessing.

Sita bent and touched her grandmother's feet, and then Grandfather pushed off. The little boat—with its two old people and three goats—riding swiftly on the river, moved slowly, very slowly, towards the opposite bank. The current was so swift now that Sita realized the boat would be carried about half a mile downstream before Grandfather could get it to dry land.

It bobbed about on the water, getting smaller and smaller, until it was just a speck on the broad river.

And suddenly Sita was alone.

There was a wind, whipping the raindrops against her face; and there was the water, rushing past the island; and there was the distant shore, blurred by rain; and there was the small hut; and there was the tree.

Sita got busy. The hens had to be fed. They weren't bothered about anything except food. Sita threw them handfuls of coarse grain and potato peelings and peanut shells.

Then she took the broom and swept out the hut, lit the charcoal burner, warmed some milk, and thought, 'Tomorrow there will be no milk...' She began peeling onions. Soon her eyes started smarting and, pausing for a few moments and glancing around the quiet room, she became aware again that she was alone. Grandfather's hookah stood by itself in one corner. It was a beautiful old hookah, which had belonged to Sita's great-grandfather. The bowl was made out of a coconut encased in silver. The long winding stem was at least four feet in length. It was their most valuable possession. Grandmother's sturdy shisham wood walking stick stood in another corner.

Sita looked around for Mumta, found the doll beneath the cot, and placed her within sight and hearing.

Thunder rolled down from the hills. BOOM—BOOM—BOOM...

'The gods of the mountains are angry,' said Sita. 'Do you think they are angry with me?'

'Why should they be angry with you?' asked Mumta.

'They don't have to have a reason for being angry. They are angry with everything, and we are in the middle of everything. We are so small—do you think they know we are here?'

'Who knows what the gods think?'

'But I made you,' said Sita, 'and I know you are here.'

'And will you save me if the river rises?'

'Yes, of course. I won't go anywhere without you, Mumta.'

Sita couldn't stay indoors for long. She went out, taking Mumta with her, and stared out across the river, to the safe land on the other side. But was it safe there? The river looked much wider now. Yes, it had crept over its banks and spread far across the flat plain. Far away, people were driving their cattle through waterlogged, flooded fields, carrying their belongings in bundles on their heads or shoulders, leaving their homes, making for the high land. It wasn't safe anywhere.

She wondered what had happened to Grandfather and Grandmother. If they had reached the shore safely, Grandfather would have to engage a bullock cart, or a pony-drawn carriage, to get Grandmother to the district town, five or six miles away, where there was a market, a court, a jail, a cinema and a hospital.

She wondered if she would ever see Grandmother again. She had done her best to look after the old lady, remembering the times when Grandmother had looked after her, had gently touched her fevered brow, and had told her stories—stories about the gods: about the young Krishna, friend of birds and animals, so full of mischief, always causing confusion among the other gods; and Indra, who made the thunder and lightning; and Vishnu, the preserver of all good things, whose steed was a great white bird; and Ganesh, with the elephant's head; and Hanuman, the monkey god, who helped the young Prince Rama in his war with the King of Ceylon. Would Grandmother return to tell her more about them, or would she have to find out for herself?

The island looked much smaller now. In parts, the mud banks had dissolved quickly, sinking into the river. But in the middle of the island there was rocky ground, and the rocks would never crumble, they could only be submerged. In a space in the middle of the rocks grew the tree.

Sita climbed up the tree to get a better view. She had climbed the tree many times and it took her only a few seconds to reach the higher branches. She put her hand to her eyes to shield them from the rain, and gazed upstream.

There was water everywhere. The world had become one vast river. Even the trees on the forested side of the river looked as though they had grown from the water, like mangroves. The sky was banked with massive, moisture-laden clouds. Thunder rolled down from the hills and the river seemed to take it up with a hollow, booming sound.

Something was floating down with the current, something big and bloated. It was closer now, and Sita could make out the bulky object—a drowned buffalo, being carried rapidly downstream.

So the water had already inundated the villages further upstream. Or perhaps the buffalo had been grazing too close to the rising river.

Sita's worst fears were confirmed when, a little later, she saw planks of wood, small trees and bushes, and then a wooden bedstead, floating past the island.

How long would it take for the river to reach her own small hut?

As she climbed down from the tree, it began to rain more heavily. She ran indoors, shooing the hens before her. They flew into the hut and huddled under Grandmother's cot. Sita thought it would be best to keep them together now. And having them with her took away some of the loneliness.

There were three hens and a cock bird. The river did not bother them. They were interested only in food, and Sita kept them happy by throwing them a handful of onion skins.

She would have liked to close the door and shut out the swish of the rain and the boom of the river, but then she would have no way of knowing how fast the water rose.

She took Mumta in her arms, and began praying for the rain to stop and the river to fall. She prayed to the god Indra, and, just in case he was busy elsewhere, she prayed to other gods too. She prayed for the safety of her grandparents and for her own safety. She put herself last but only with great difficulty.

She would have to make herself a meal. So she chopped

up some onions, fried them, then added turmeric and red chilli powder and stirred until she had everything sizzling; then she added a tumbler of water, some salt, and a cup of one of the cheaper lentils. She covered the pot and allowed the mixture to simmer. Doing this took Sita about ten minutes. It would take at least half an hour for the dish to be ready.

When she looked outside, she saw pools of water amongst the rocks and near the tree. She couldn't tell if it was rainwater or overflow from the river.

She had an idea.

A big tin trunk stood in a corner of the room. It had belonged to Sita's mother. There was nothing in it except a cotton-filled quilt, for use during the cold weather. She would stuff the trunk with everything useful or valuable, and weigh it down so that it wouldn't be carried away—just in case the river came over the island… Grandfather's hookah went into the trunk. Grandmother's walking stick went in too. So did a number of small tins containing the spices used in cooking—nutmeg, caraway seeds, cinnamon, coriander and pepper—a bigger tin of flour and a tin of raw sugar. Even if Sita had to spend several hours in the tree, there would be something to eat when she came down again.

A clean white cotton shirt of Grandfather's, and Grandmother's only spare sari also went into the trunk. Never mind if they got stained with yellow curry powder! Never mind if they got to smell of salted fish, some of that went in too.

Sita was so busy packing the trunk that she paid no attention to the lick of cold water at her heels. She locked the trunk, placed the key high on the rock wall, and turned to give her attention to the lentils. It was only then that she discovered that she was walking about on a watery floor.

She stood still, horrified by what she saw. The water was

oozing over the threshold, pushing its way into the room.

Sita was filled with panic. She forgot about her meal and everything else. Darting out of the hut, she ran splashing through ankle-deep water towards the safety of the peepul tree. If the tree hadn't been there, such a well-known landmark, she might have floundered into deep water, into the river.

She climbed swiftly into the strong arms of the tree, made herself secure on a familiar branch and thrust the wet hair away from her eyes.

◆

She was glad she had hurried. The hut was now surrounded by water. Only the higher parts of the island could still be seen—a few rocks, the big rock on which the hut was built, a hillock on which some thorny bilberry bushes grew.

The hens hadn't bothered to leave the hut. They were probably perched on the cot now.

Would the river rise still higher? Sita had never seen it like this before. It swirled around her, stretching in all directions.

More drowned cattle came floating down. The most unusual things went by on the water—an aluminium kettle, a cane chair, a tin of tooth powder, an empty cigarette packet, a wooden slipper, a plastic doll...

A doll!

With a sinking feeling, Sita remembered Mumta.

Poor Mumta! She had been left behind in the hut. Sita, in her hurry, had forgotten her only companion.

Well, thought Sita, if I can be careless with someone I've made, how can I expect the gods to notice me, alone in the middle of the river?

The waters were higher now, the island fast disappearing.

Something came floating out of the hut.

It was an empty kerosene tin, with one of the hens perched on top. The tin came bobbing along on the water, not far from the tree, and was then caught by the current and swept into the river. The hen still managed to keep its perch.

A little later, the water must have reached the cot because the remaining hens flew up to the rock ledge and sat huddled there in the small recess.

The water was rising rapidly now, and all that remained of the island was the big rock that supported the hut, the top of the hut itself and the peepul tree.

It was a tall tree with many branches and it seemed unlikely that the water could ever go right over it. But how long would Sita have to remain there? She climbed a little higher, and as she did so, a jet-black jungle crow settled in the upper branches, and Sita saw that there was a nest in them—a crow's nest, an untidy platform of twigs wedged in the fork of a branch.

In the nest were four blue-green, speckled eggs. The crow sat on them and cawed disconsolately. But though the crow was miserable, its presence brought some cheer to Sita. At least she was not alone. Better to have a crow for company than no one at all.

Other things came floating out of the hut—a large pumpkin; a red turban belonging to Grandfather, unwinding in the water like a long snake; and then—Mumta! The doll, being filled with straw and wood shavings, moved quite swiftly on the water and passed close to the peepul tree. Sita saw it and wanted to call out, to urge her friend to make for the tree, but she knew that Mumta could not swim—the doll could only float, travel with the river and perhaps be washed ashore many miles downstream.

The tree shook in the wind and the rain. The crow cawed and flew up, circled the tree a few times and returned to the nest. Sita clung to her branch.

The tree trembled throughout its tall frame. To Sita it felt like an earthquake tremor; she felt the shudder of the tree in her own bones.

The river swirled all around her now. It was almost up to the roof of the hut. Soon the mud walls would crumble and vanish. Except for the big rock and some trees far, far away, there was only water to be seen.

For a moment or two Sita glimpsed a boat with several people in it moving sluggishly away from the ruins of a flooded village, and she thought she saw someone pointing towards her, but the river swept them on, and the boat was lost to view.

The river was very angry; it was like a wild beast, a dragon on the rampage, thundering down from the hills, and sweeping across the plain, bringing with it dead animals, uprooted trees, household goods and huge fish choked to death by the swirling mud.

The tall, old peepul tree groaned. Its long, winding roots clung tenaciously to the earth from which the tree had sprung many, many years ago. But the earth was softening; the stones were being washed away. The roots of the tree were rapidly losing their hold.

The crow must have known that something was wrong, because it kept flying up and circling the tree, reluctant to settle in it and reluctant to fly away. As long as the nest was there, the crow would remain, flapping about and cawing in alarm.

Sita's wet cotton dress clung to her thin body. The rain ran down from her long black hair. It poured from every leaf of the tree. The crow, too, was drenched and groggy.

The tree groaned and moved again. It had seen many monsoons. Once before, it had stood firm while the river had swirled around its massive trunk. But it had been young then. Now, old in years and tired of standing still, the tree was ready to join the river.

With a flurry of its beautiful leaves, and a surge of mud from below, the tree left its place in the earth, and, tilting, moved slowly forward, turning a little from side to side, dragging its roots along the ground. To Sita, it seemed as though the river was rising to meet the sky.

Then the tree moved into the main current of the river, and went a little faster, swinging Sita from side to side. Her feet were in the water but she clung tenaciously to her branch.

◆

The branches swayed, but Sita did not lose her grip. The water was very close now. Sita was frightened. She could not see the extent of the flood or the width of the river. She could only see the immediate danger—the water surrounding the tree.

The crow kept flying around the tree. The bird was in a terrible rage. The nest was still in the branches, but not for long... The tree lurched and twisted slightly to one side, and the nest fell into the water. Sita saw the eggs go one by one.

The crow swooped low over the water, but there was nothing it could do. In a few moments, the nest had disappeared.

The bird followed the tree for about fifty yards, as though hoping that something still remained in the tree. Then, flapping its wings, it rose high into the air and flew across the river until it was out of sight.

Sita was alone once more. But there was no time for feeling lonely. Everything was in motion—up and down and sideways and forwards. 'Any moment,' thought Sita, 'the tree will turn right over and I'll be in the water!'

She saw a turtle swimming past—a great river turtle, the kind that feeds on decaying flesh. Sita turned her face away. In the distance she saw a flooded village and people in flat-bottomed boats but they were very far away.

Because of its great size, the tree did not move very swiftly on the river. Sometimes, when it passed into shallow water, it stopped, its roots catching in the rocks; but not for long—the river's momentum soon swept it on. At one place, where there was a bend in the river, the tree struck a sandbank and was still.

Sita felt very tired. Her arms were aching and she was no longer upright. With the tree almost on its side, she had to cling tightly to her branch to avoid falling off.

The grey weeping sky was like a great shifting dome. She knew she could not remain much longer in that position. It might be better to try swimming to some distant rooftop or tree. Then she heard someone calling.

Craning her neck to look upriver, she was able to make out a small boat coming directly towards her.

The boat approached the tree. There was a boy in the boat who held on to one of the branches to steady himself, giving his free hand to Sita. She grasped it, and slipped into the boat beside him. The boy placed his bare foot against the tree trunk and pushed away. The little boat moved swiftly down the river. The big tree was left far behind. Sita would never see it again.

◆

She lay stretched out in the boat, too frightened to talk. The boy looked at her, but he did not say anything, he did not even smile. He lay on his two small oars, stroking smoothly, rhythmically, trying to keep from going into the middle of the river. He wasn't strong enough to get the boat right out of the swift current, but he kept trying.

A small boat on a big river—a river that had no boundaries but that reached across the plains in all directions. The boat moved swiftly on the wild waters, and Sita's home was left far behind.

The boy wore only a loincloth. A sheathed knife was knotted into his waistband. He was a slim, wiry boy, with a hard, flat belly; he had high cheekbones, strong white teeth. He was a little darker than Sita.

'You live on the island,' he said at last, resting on his oars, and allowing the boat to drift a little, for he had reached a broader, more placid stretch of the river.

'I have seen you sometimes. But where are the others?'

'My grandmother was sick,' said Sita, 'so Grandfather took her to the hospital in Shahganj.'

'When did they leave?'

'Early this morning.'

Only that morning—and yet it seemed to Sita as though it had been many mornings ago.

'Where have you come from?' she asked. She had never seen the boy before.

'I come from...' he hesitated, '...near the foothills. I was in my boat, trying to get across the river with the news that one of the villages was badly flooded, but the current was too strong. I was swept down past your island. We cannot fight the river, we must go wherever it takes us.'

'You must be tired. Give me the oars.'

'No. There is not much to do now, except keep the boat steady.'

He brought in one oar, and with his free hand he felt under the seat where there was a small basket. He produced two mangoes, and gave one to Sita.

They bit deep into the ripe, fleshy mangoes, using their teeth to tear the skin away. The sweet juice trickled down their chins. The flavour of the fruit was heavenly—truly this was the nectar of the gods!

Sita hadn't tasted a mango for over a year. For a few moments she forgot about the flood—all that mattered was the mango!

The boat drifted, but not so swiftly now, for as they went further away across the plains, the river lost much of its tremendous force.

'My name is Krishan,' said the boy. 'My father has many cows and buffaloes, but several have been lost in the flood.'

'I suppose you go to school,' said Sita.

'Yes, I am supposed to go to school. There is one not far from our village. Do you have to go to school?'

'No—there is too much work at home.'

It was no use wishing she was at home—home wouldn't be there any more—but she wished, at that moment, that she had another mango.

Towards evening, the river changed colour. The sun, low in the sky, emerged from behind the clouds, and the river changed slowly from grey to gold, from gold to a deep orange, and then, as the sun went down, all these colours were drowned in the river, and the river took on the colour of the night.

The moon was almost at the full and Sita could see across the river, to where the trees grew on its banks.

'I will try to reach the trees,' said the boy, Krishan. 'We do not want to spend the night on the water, do we?'

And so he pulled for the trees. After ten minutes of strenuous rowing, he reached a turn in the river and was able to escape the pull of the main current. Soon they were in a forest, rowing between tall evergreens.

◆

They moved slowly now, paddling between the trees, and the moon lighted their way, making a crooked silver path over the water.

'We will tie the boat to one of these trees,' said Krishan. 'Then we can rest. Tomorrow we will have to find our way out of the forest.'

He produced a length of rope from the bottom of the boat, tied one end to the boat's stern, and threw the other end over a stout branch which hung only a few feet above the water. The boat came to rest against the trunk of the tree.

It was a tall, sturdy toon tree—the Indian mahogany—and it was quite safe, for there was no rush of water here; besides, the trees grew close together, making the earth firm and unyielding.

But the denizens of the forest were on the move.

The animals had been flooded out of their holes, caves and lairs, and were looking for shelter and dry ground.

Sita and Krishan had barely finished tying the boat to the tree when they saw a huge python gliding over the water towards them. Sita was afraid that it might try to get into the boat; but it went past them, its head above water, its great awesome length trailing behind, until it was lost in the shadows.

Krishan had more mangoes in the basket, and he and Sita sucked hungrily on them while they sat in the boat.

A big sambar stag came thrashing through the water. He did not have to swim; he was so tall that his head and shoulders remained well above the water. His antlers were big and beautiful.

'There will be other animals,' said Sita. 'Should we climb into the tree?'

'We are quite safe in the boat,' said Krishan. 'The animals are interested only in reaching dry land. They will not even hunt each other. Tonight, the deer are safe from the panther and the tiger. So lie down and sleep, and I will keep watch.'

Sita stretched herself out in the boat and closed her eyes, and the sound of the water lapping against the sides of the boat soon lulled her to sleep. She woke once, when a strange bird called overhead. She raised herself on one elbow, but Krishan was awake, sitting in the prow, and he smiled reassuringly at her.

He looked blue in the moonlight, the colour of the young

god Krishna, and for a few moments Sita was confused and wondered if the boy was indeed Krishna; but when she thought about it, she decided that it wasn't possible. He was just a village boy and she had seen hundreds like him—well, not exactly like him; he was different, in a way she couldn't explain to herself...

And when she slept again, she dreamt that the boy and Krishna were one, and that she was sitting beside him on a great white bird which flew over mountains, over the snow peaks of the Himalaya, into the cloudland of the gods. There was a great rumbling sound, as though the gods were angry about the whole thing, and she woke up to this terrible sound and looked about her, and there in the moonlit glade, up to his belly in water, stood a young elephant, his trunk raised as he trumpeted his predicament to the forest—for he was a young elephant, and he was lost, and he was looking for his mother.

He trumpeted again, and then lowered his head and listened. And presently, from far away, came the shrill trumpeting of another elephant. It must have been the young one's mother, because he gave several excited trumpet calls, and then went stamping and churning through the flood water towards a gap in the trees. The boat rocked in the waves made by his passing.

'It's all right now,' said Krishan. 'You can go to sleep again.'

'I don't think I will sleep now,' said Sita.

'Then I will play my flute for you,' said the boy, 'and the time will pass more quickly.'

From the bottom of the boat he took a flute, and putting it to his lips, he began to play. The sweetest music that Sita had ever heard came pouring from the little flute, and it seemed to fill the forest with its beautiful sound. And the music carried her away again, into the land of dreams, and they were riding on the bird once more, Sita and the blue god, and they were passing through clouds and mist, until suddenly the sun shot out through the clouds. And at the same moment, Sita opened her eyes and saw the sun streaming through the branches of the toon tree, its bright green leaves making a dark pattern against the blinding blue of the sky.

Sita sat up with a start, rocking the boat. There were hardly any clouds left. The trees were drenched with sunshine.

The boy Krishan was fast asleep at the bottom of the boat. His flute lay in the palm of his half-opened hand. The sun came slanting across his bare brown legs. A leaf had fallen on his upturned face, but it had not woken him; it lay on his cheek as though it had grown there.

Sita did not move again. She did not want to wake the boy. It didn't look as though the water had gone down, but it hadn't risen, and that meant the flood had spent itself.

The warmth of the sun, as it crept up Krishan's body, woke him at last. He yawned, stretched his limbs, and sat up beside Sita.

'I'm hungry,' he said with a smile.

'So am I,' said Sita.

'The last mangoes,' he said, and emptied the basket of its last two mangoes.

After they had finished the fruit, they sucked the big seeds until they were quite dry. The discarded seeds floated well on the water. Sita had always preferred them to paper boats.

'We had better move on,' said Krishan.

He rowed the boat through the trees, and then for about an hour they were passing through the flooded forest, under the dripping branches of rain-washed trees.

Sometimes they had to use the oars to push away vines and creepers. Sometimes drowned bushes hampered them. But they were out of the forest before noon.

Now the water was not very deep and they were gliding over flooded fields. In the distance they saw a village. It was on high ground. In the old days, people had built their villages on hilltops, which gave them a better defence against bandits and invading armies.

This was an old village, and though its inhabitants had long ago exchanged their swords for pruning forks, the hill on which it stood now protected it from the flood.

The people of the village—long-limbed, sturdy Jats—were generous, and gave the stranded children food and shelter. Sita was anxious to find her grandparents, and an old farmer who had business in Shahganj offered to take her there. She was hoping that Krishan would accompany her, but he said he would wait in the village, where he knew others would soon be arriving, his own people among them.

'You will be all right now,' said Krishan. 'Your grandfather will be anxious for you, so it is best that you go to him as soon as you can. And in two or three days, the water will go down and you will be able to return to the island.'

'Perhaps the island has gone forever,' said Sita.

As she climbed into the farmer's bullock cart, Krishan handed her his flute.

'Please keep it for me,' he said. 'I will come for it one day.'

And when he saw her hesitate, he added, his eyes twinkling, 'It is a good flute!'

◆

It was slow going in the bullock cart. The road was awash, the wheels got stuck in the mud, and the farmer, his grown son and Sita had to keep getting down to heave and push in order to free the big, wooden wheels.

They were still in a foot or two of water. The bullocks were bespattered with mud, and Sita's legs were caked with it. They were a day and a night in the bullock cart before they reached Shahganj; by that time, Sita, walking down the narrow bazaar of the busy market town, was hardly recognizable.

Grandfather did not recognize her. He was walking stiffly down the road, looking straight ahead of him, and would have walked right past the dusty, dishevelled girl if she had not charged straight at his thin, shaky legs and clasped him around the waist.

'Sita!' he cried, when he had recovered his wind and his balance.

'But how are you here? How did you get off the island? I was so worried—it has been very bad these last two days...'

'Is Grandmother all right?' asked Sita.

But even as she spoke, she knew that Grandmother was no longer with them. The dazed look in the old man's eyes told her as much. She wanted to cry, not for Grandmother, who could suffer no more, but for Grandfather, who looked so helpless and bewildered; she did not want him to be unhappy. She forced back her tears, took his gnarled and trembling hand, and led him down the crowded street. And she knew, then, that it would be on her shoulder that Grandfather would have to lean in the years to come.

They returned to the island after a few days, when the river was no longer in spate. There was more rain, but the worst was over. Grandfather still had two of the goats; it had not been necessary to sell more than one.

He could hardly believe his eyes when he saw that the tree had disappeared from the island—the tree that had seemed as permanent as the island, as much a part of his life as the river itself. He marvelled at Sita's escape.

'It was the tree that saved you,' he said.

'And the boy,' said Sita.

Yes, and the boy.

She thought about the boy, and wondered if she would ever see him again. But she did not think too much, because there was so much to do.

For three nights they slept under a crude shelter made out of jute bags. During the day she helped Grandfather rebuild the mud hut. Once again, they used the big rock as a support.

The trunk, which Sita had packed so carefully, had not been swept off the island, but the water had got into it, and the food and clothing had been spoilt. But Grandfather's hookah had been saved, and, in the evenings, after their work was done and they had eaten the light meal that Sita prepared, he would smoke with a little of his old contentment, and tell Sita about other floods and storms that he had experienced as a boy.

Sita planted a mango seed in the same spot where the peepul tree had stood. It would be many years before it grew into a big tree, but Sita liked to imagine sitting in its branches one day, picking the mangoes straight from the tree and feasting on them all day. Grandfather was more particular about making a vegetable garden and putting down peas, carrots, gram and mustard.

One day, when most of the hard work had been done and the new hut was almost ready, Sita took the flute that had been given to her by the boy, and walked down to the water's edge and tried to play it.

But all she could produce were a few broken notes, and even the goats paid no attention to her music.

Sometimes Sita thought she saw a boat coming down the river and she would run to meet it; but usually there was no boat, or if there was, it belonged to a stranger or to another fisherman. And so she stopped looking out for boats. Sometimes she thought she heard the music of a flute, but it seemed very distant and she could never tell where the music came from.

Slowly, the rains came to an end. The flood waters had receded, and in the villages people were beginning to till the land again and sow crops for the winter months. There were cattle fairs and wrestling matches. The days were warm and sultry. The water in the river was no longer muddy, and one evening Grandfather brought home a huge mahseer fish and Sita made it into a delicious curry.

◆

Grandfather sat outside the hut, smoking his hookah. Sita was at the far end of the island, spreading clothes on the rocks to dry.

One of the goats had followed her. It was the friendlier of the two, and often followed Sita about the island. She had made it a necklace of coloured beads.

She sat down on a smooth rock, and, as she did so, she noticed a small, bright object in the sand near her feet. She stooped and picked it up. It was a little wooden toy—a coloured peacock—that must have come down on the river and been swept ashore on the island. Some of the paint had rubbed off, but for Sita, who had no toys, it was a great find. Perhaps it would speak to her, as Mumta had spoken to her.

As she held the toy peacock in the palm of her hand, she thought she heard the flute music again, but she did not look up. She had heard it before, and she was sure that it was all in her mind. But this time the music sounded nearer, much nearer. There was a soft footfall in the sand. And, looking up, she saw the boy, Krishan, standing over her.

'I thought you would never come,' said Sita.

'I had to wait until the rains were over. Now that I am free, I will come more often. Did you keep my flute?'

'Yes, but I cannot play it properly. Sometimes it plays by itself, I think, but it will not play for me!'

'I will teach you to play it,' said Krishan.

He sat down beside her, and they cooled their feet in the water, which was clear now, reflecting the blue of the sky. You could see the sand and the pebbles of the riverbed.

'Sometimes the river is angry, and sometimes it is kind,' said Sita.

'We are part of the river,' said the boy. 'We cannot live without it.'

It was a good river, deep and strong, beginning in the mountains and ending in the sea. Along its banks, for hundreds of miles, lived millions of people, and Sita was only one small girl among them, and no one had ever heard of her, no one knew her—except for the old man, the boy and the river.

THE BANYAN TREE

Though the house and grounds belonged to my grandparents, the magnificent old banyan tree was mine—chiefly because Grandfather, at sixty-five, could no longer climb it.

Its spreading branches, which hung to the ground and took root again, forming a number of twisting passages, gave me endless pleasure. Among them were squirrels and snails and butterflies. The tree was older than the house, older than Grandfather, as old as Dehradun itself. I could hide myself in its branches, behind thick green leaves, and spy on the world below.

My first friend was a small grey squirrel. Arching his back and sniffing into the air, he seemed at first to resent my invasion of his privacy. But when he found that I did not arm myself with catapult or airgun, he became friendly, and when I started bringing him pieces of cake and biscuit, he grew quite bold and was soon taking morsels from my hand.

Before long he was delving into my pockets and helping himself to whatever he could find. He was a very young squirrel and his friends and relatives probably thought him foolish and headstrong for trusting a human.

In the spring, when the banyan tree was full of small red figs, birds of all kinds would flock into its branches: the red-bottomed bulbul, cheerful and greedy; gossipy rosy pastors; parrots, mynahs and crows squabbling with one another. During the fig season, the banyan tree was the noisiest place in the garden.

Halfway up the tree I had built a crude platform where I would spend the afternoons when it was not too hot. I could read there, propping myself up against the bole of the tree with a cushion from the living room. *Treasure Island*, *The Adventures of Huckleberry Finn* and *The Story of Doctor Dolittle* were some of the books that made up my banyan tree library.

When I did not feel like reading, I could look down through the leaves at the world below. And, on one particular afternoon, I had a grandstand view of that classic of the Indian wilds, a fight between a mongoose and a cobra. And this one had not been staged for my benefit!

The warm breezes of approaching summer had sent everyone, including the gardener, into the house. I was feeling drowsy myself, wondering if I should go to the pond and have a swim with Ramu and the buffaloes, when I saw a huge black cobra gliding out of a clump of cactus. At the same time a mongoose emerged from the bushes and went straight for the cobra.

In a clearing beneath the banyan tree, in bright sunshine, they came face to face.

The cobra knew only too well that the grey mongoose, three feet long, was a superb fighter, clever and aggressive. But the cobra, too, was a skilful and experienced fighter. He could move swiftly and strike with the speed of light; and the sacs behind his long, sharp fangs were full of deadly poison.

It was to be a battle of champions.

Hissing defiance, his forked tongue darting in and out, the cobra raised three of his six feet off the ground and spread his spectacled hood. The mongoose bushed his tail. The long hair on his spine stood up.

Though the combatants were unaware of my presence in the tree, they were soon made aware of the arrival of two other spectators. One was a mynah, the other a jungle crow. They had seen these preparations for battle and had settled on the cactus to watch the outcome. Had they been content only to watch, all would have been well with both of them.

The cobra stood on the defensive, swaying slowly from side to side, trying to mesmerize the mongoose into making a false move. But the mongoose knew the power of his opponent's glassy, unwinking eyes, and refused to meet them. Instead, he fixed his gaze at a point just below the cobra's hood and opened the attack.

Moving forward quickly, until he was just within the cobra's reach, the mongoose made a pretended move to one side. Immediately the cobra struck. His great hood came down so swiftly that I thought nothing could save the mongoose. But the little fellow jumped neatly to one side, and darted in as swiftly as the cobra, biting the snake on the back and darting away again out of reach.

At the same moment that the cobra struck, the crow and the mynah hurled themselves at him, only to collide heavily in mid-air. Shrieking insults at each other, they returned to the cactus plant.

A few drops of blood glistened on the cobra's back. The cobra struck again and missed. Again the mongoose sprang aside, jumped in and bit. Again, the birds dived at the snake, bumped into each other instead and returned shrieking to the safety of the cactus.

The third round followed the same course as the first two with one dramatic difference. The crow and the mynah, determined to take part in the proceedings, dived at the snake, but this time they missed each other as well as their mark. The mynah flew on and reached its perch, but the crow tried to pull up in mid-air and turn back. In the second that it took the bird to do this, the cobra whipped his head back and struck with great force, his snout thudding against the crow's body.

I saw the bird flung nearly twenty feet across the garden. It fluttered about for a while, then lay still. The mynah remained on the cactus plant, and when the snake and the mongoose returned to the fight, very wisely decided not to interfere again!

The cobra was weakening, and the mongoose, walking fearlessly up to it, raised himself on his short legs and with a lightning snap had the big snake by the snout. The cobra writhed and lashed about in a frightening manner, and even coiled itself about the mongoose, but to no avail. The little fellow hung grimly on, until the snake had ceased to struggle. He then smelt it along its quivering length, gripped it round the hood and dragged it into the bushes.

The mynah dropped cautiously to the ground, hopped about, jeered into the bushes from a safe distance, and then, with a shrill cry of congratulation, flew away.

The banyan tree was also the setting for what we were to call the Strange Case of the Grey Squirrel and the White Rat.

The white rat was Grandfather's—he had bought it for one-quarter of a rupee but I would often take it with me into the banyan tree, where it soon struck up a friendship with one of the squirrels. They would go off together on little excursions among the roots and branches of the old tree.

Then the squirrel started building a nest. At first she tried building it in my pockets, and when I went indoors and took off my clothes I would find straw and grass falling out.

Then one day Grandmother's knitting was missing. We hunted for it everywhere but without success.

The next day I saw something glinting in a hole in the tree. Going up to investigate, I saw that it was the end of Grandmother's steel knitting needle. On looking further, I discovered that the hole was crammed with knitting. Amongst the wool were three baby squirrels—and all of them were white!

We gazed at the white squirrels in wonder and fascination. Grandfather was puzzled at first, but when I told him about the white rat's visits to the tree, his brow cleared. He said the white rat must be the father.

THE CORAL TREE

The night had been hot, the rain frequent, and I had been sleeping on the verandah instead of in the house. I was in my twenties, had begun to earn a living and felt I had certain responsibilities.

In a short time, a tonga would take me to the railway station, and from there a train would take me to Bombay, and then a ship would take me to England. There would be work, interviews, a job, a different kind of life, so many things that this small bungalow of my grandfather would be remembered fitfully, in rare moments of reflection.

When I awoke on the verandah, I saw a grey morning, smelt the rain on the red earth and remembered that I had to go away. A girl was standing on the verandah porch, looking at me very seriously. When I saw her, I sat up in bed with a start.

She was a small, dark girl, her eyes big and black, her pigtails tied up in a bright red ribbon, and she was fresh and clean like the rain and the red earth.

She stood looking at me and was very serious.

'Hello,' I said, smiling and trying to put her at ease. But the girl was business-like and acknowledged my greeting with a brief nod.

'Can I do anything for you?' I asked, stretching my limbs. 'Do you stay nearby?'

With great assurance she said, 'Yes, but I can stay on my own.'

'You're like me,' I said, and for a while, forgot about being an old man of twenty. 'I like to be on my own but I'm going away today.'

'Oh,' she said, a little breathlessly.

'Would you care to go to England?'

'I want to go everywhere,' she said. 'To America and Africa and Japan and Honolulu.'

'Maybe you will,' I said. 'I'm going everywhere, and no one can stop me... But what is it you want, what did you come for?'

'I want some flowers but I can't reach them.' She waved her hand towards the garden, 'That tree, see?'

The coral tree stood in front of the house surrounded by pools of water and broken, fallen blossoms. The branches of the tree were thick with scarlet, pea-shaped flowers.

'All right, just let me get ready.'

The tree was easy to climb and I made myself comfortable on one of the lower branches, smiling down at the serious, upturned face of the girl.

'I'll throw them down to you,' I said.

I bent a branch but the wood was young and green and I had to twist it several times before it snapped.

'I'm not sure I ought to do this,' I said as I dropped the flowering branch to the girl.

'Don't worry,' she said.

I felt a sudden nostalgic longing for childhood and an urge to remain behind in my grandfather's house with its tangled memories and ghosts of yesteryear. But I was the only one left, and what could I do except climb tamarind and jackfruit trees?

'Have you many friends?' I asked.

'Oh yes.'

'And who is the best?'

'The cook. He lets me stay in the kitchen, which is more interesting than the house. And I like to watch him cooking. And he gives me things to eat and tells me stories...'

'And who is your second best friend?'

She inclined her head to one side and thought very hard. 'I'll make you second best,' she said.

I sprinkled coral blossoms on her head. 'That's very kind of you. I'm happy to be second best.'

A tonga bell sounded at the gate and I looked out from the tree and said, 'It's come for me. I have to go now.'

I climbed down.

'Will you help me with my suitcases?' I asked, as we walked together towards the verandah. 'There's no one here to help me. I am the last to go. Not because I want to go but because I have to.'

I sat down on the cot and packed a few last things in my suitcase. All the doors of the house were locked. On my way to the station, I would leave the keys with the caretaker. I had already given instructions to the agent to try and sell the house. There was nothing more to be done. We walked in silence to the waiting tonga, thinking and wondering about each other. The girl stood at the side of the path, on the damp earth, looking at me.

'Thank you,' I said, 'I hope I shall see you again.'

'I'll see you in London,' she said. 'Or America or Japan, I want to go everywhere.'

'I'm sure you will,' I said. 'And perhaps I'll come back and we'll meet again in this garden. That would be nice, wouldn't it?'

She nodded and smiled. We knew it was an important moment. The tonga driver spoke to his pony and the carriage set off down the gravel path, rattling a little. The girl and I waved to each other. In the girl's hand was a sprig of coral blossom. As she waved, the blossoms fell apart and danced lightly in the breeze.

'Goodbye!' I called.

'Goodbye!' called the girl.

The ribbon had come loose from her pigtail and lay on the ground with the coral blossoms.

And she was fresh and clean like the rain and the red earth.

RHODODENDRONS IN THE MIST

Blood-red, the fallen blossoms lay on the snow, even more striking when laid bare. On the trees they blended with the foliage. On the ground, on those patches of recent snow, they seemed to be bleeding.

It had been a harsh winter in the hills, and it was still snowing at the end of March. But this was flowering time for the rhododendron trees, and they blossomed in sun, snow, or pelting rain. By mid-afternoon the hill station was shrouded in a heavy mist, and the trees stood out like ghostly sentinels.

The hill station wasn't Simla, where I had gone to school, or Mussoorie, where I was to settle later on. It was Dalhousie, a neglected and almost forgotten hill station in the western Himalaya. But Dalhousie had the best rhododendron trees, and they grew all over the mountain, showing off before the colourless oaks and drooping pines.

But I wasn't in Dalhousie for the rhododendrons. It was 1959, and the Dalai Lama had just fled from Tibet, seeking sanctuary in India. Thousands of his followers and fellow-Tibetans had fled with him, and these refugees had to be settled somewhere. Dalhousie, with its many empty houses, was ideal for this purpose, and a carpet-weaving centre had been set up on one of the estates. The Tibetans made beautiful rugs and carpets. I know nothing about carpet-weaving, but I was working for CARE, an American relief organization, and I had been sent to Dalhousie (with the approval of the Government of India) to assess the needs of the refugees.

This is not the story of my tryst with the Tibetans, although

I did suffer greatly from drinking large quantities of butter tea, which travels very slowly down the gullet and feels like lead by the time it reaches your stomach. The carpet-weaving centre became a great success, and I went on to work for CARE for several years; but that's another story. Out of one experience came another experience, as often happens during our peregrinations on planet Earth, and it was during my stay in Dalhousie that I had a strange and rather unsettling experience.

I was staying at a small hotel that was quite empty as no one visited Dalhousie in those days and certainly not at the end of March. The hill station had been convenient for visitors from Lahore, but Partition had put an end to that.

◆

The hotel had a small garden, bare at this time of the year. But on the second day of my stay, returning from the carpet-weaving centre, I noticed that there was a gardener working on the flower beds, digging around and transplanting some seedlings. He looked up as I passed, and for a moment I thought I knew him. There was something familiar about his features—the slit eyes, the broad, flattened nose, the harelip—yes, the cleft lip was very noticeable—but he wasn't anyone I knew or had known. At least I didn't think so... He was just a likeness to someone I had seen somehow, somewhere else. It was a bit of a tease.

And it would have remained just that if he hadn't looked up and met my gaze.

A flood of recognition crossed his face. But then he looked away, almost as though he did not want to recognize me; or be recognized.

I passed him. It was curious, but it didn't bother me. We keep bumping into people who look slightly familiar. It is said that everyone has a double somewhere on this planet. I had yet to meet mine—God forbid!—but perhaps I was seeing someone else's double.

I was relaxing in the verandah later that evening, browsing through an old magazine, when the gardener passed me on his way to the garden shed to put away his tools. There was something about his walk that brought back an image from the past. He had a slight limp. And when he looked at me again, his harelip registered itself on my memory. And now I recognized him. And of course he knew me.

I was the man who'd caught him rifling through my landlady's cupboards and drawers in Dehradun, some three years previously. I had exposed him, reported him, suggested she dismiss him; but the old lady, a widow, had grown quite fond of the youth, and had kept him in her service. He was good at running about and making himself useful, and, in spite of his cleft lip, he was not unattractive.

When I left Dehradun to take up my job in Delhi, I had forgotten the matter, almost forgotten the young man and my landlady; it was another tenant who informed me that the youth—his name was Sohan—had stabbed the old lady and made off with the contents of her jewel case and other valuables. She had died in hospital a few days later.

Sohan hadn't been caught. He had obviously left the town and taken to the hills or a large city. The police had made sporadic attempts to locate him, but as time passed the case lost its urgency. The victim was not a person of importance. The criminal was a stranger, a shadowy figure of no known background.

But here he was three years later, staring me in the face. What was I to do about him? Or what was he to do about me?

◆

After Sohan had gone to his quarters, somewhere behind the hotel, I went in search of the manager. I would tell him what I knew and together we could decide on a course of action. But he had gone to a marriage and would be back late. The hotel was in charge of the cook who, a little drunk, served dinner in a hurry and retired to his quarters. 'Don't you have a night-watchman?' I asked him before he took off. 'Yes, of course,' he replied, 'Sohan, the gardener. He's the chowkidar too!'

An early retirement seemed the best thing all round, especially as I had to leave the next day. So I went to my room and made sure all the doors and windows were locked. I pushed the inside bolts all the way. I made sure the antiquated window frames were locked. As I peered out of the window, I noticed that a heavy mist had descended on the hillside. The trees stood out like ghostly apparitions, here and there a rhododendron glowing like the embers of a small fire. Then darkness enveloped the hillside. I felt cold, and wondered how much of it was fear.

I went to the bathroom and bolted the back door. Now no one could get in. Even so, I felt uneasy. Sohan was still a fugitive from the law, I had recognized him, and I was a threat to his freedom. He had killed once—perhaps more than once—and he could kill again.

I read for some time, then put out the light and tried to sleep. From a distance came the strains of music from a wedding band. Someone knocked on the door. I switched on the light and looked at my watch. It was only 10.00 p.m. Perhaps the manager had returned.

There was another knock, and I went to the door and was about to open it when some childhood words of warning from my grandmother came to mind: 'Never open the door unless you know who's there!'

'Who's there?' I called.

No answer. Just another knock.

'Who's there?' I called again.

There was a cough, a double-rap on the door.

'I'm sleeping,' I said. 'Come in the morning.' And I returned to my bed. The knocking continued but I ignored it, and after some time the person went away.

I slept a little. A couple of hours must have passed when I was woken by further knocking. But it did not come from the door. It was above me, high up on the wall. I'd forgotten there was a skylight.

I switched on the light and looked up. A face was outlined against the glass of the skylight. I could make out the flat rounded face and the harelip. It appeared to be grinning at me—rather like the disembodied head of the Cheshire cat in *Alice in Wonderland*.

The skylight was very small and I knew he couldn't crawl through the opening. But he could show me a knife—and that was what he did. It was a small clasp knife and he held it between his teeth as he peered down at me. I felt very vulnerable on the bed. So I switched off the light and moved to an old sofa at the far end of the room, where I couldn't be seen. There didn't seem to be any point in shouting for help. So I just sat there, waiting... And presumably, without a sound, he slipped away, and I remained on the sofa until the first glimmer of dawn penetrated the drawn window curtains.

◆

The manager was apologetic. 'You should have rung the bell,' he said, 'someone would have come.'

'The bell doesn't work. And someone did come...'

'I'm sorry, I'm sorry. The fellow's a villain, no doubt about it. And he's missing this morning. Your presence here must have frightened him off. So he's wanted for theft and murder. Well, we shall inform the police. Perhaps they can pick him up before he leaves the town.'

And we did inform the police. But Sohan had already taken off. The milkman had seen him boarding the early morning bus to Pathankot.

Pathankot was a busy little town on the plain below Dehradun. From there one road goes to Jammu, another to Dharamsala, a narrow-gauge railway to Kangra, and the main railway to Amritsar or Delhi. Sohan could have taken any of those routes. And no one was going to go looking for him. A police alert would be put out—a mere formality. He wasn't on their list of current criminals.

That afternoon I took a taxi to Pathankot and whiled away the evening at the railway station. My train, an overnight express to Delhi, left at 8.00 p.m. There was no rush at that time of the year. I had a first-class compartment to myself.

In those days our trains were somewhat different from what they are today. A first, second or third class compartment was usually a single carriage, or bogey. We did not have corridor trains. Bogeys were connected by steel couplings, otherwise you were not connected in any way to the other compartments. But there was an emergency cord above the upper berths, and if you pulled it, the train might stop. There were always troublemakers on the trains, just as there are today, and sometimes the chain was pulled out of mischief. As a result it was often ignored.

As the train began moving out of the station I went to all the windows and made sure that they were fastened. Then I bolted the carriage door. I was becoming adept at bolting doors and windows. Sohan was probably hiding out in some distant town or village, but I wasn't taking any chances.

The train gathered speed. The lights of Pathankot receded as we plunged into a dark and moonless night. I had a pillow and a blanket with me, and I stretched out on one of the bunks and tried to think about pleasant things such as scarlet geraniums, fragrant sweet peas, and the beautiful Nimmi, star of the silver screen; but instead I kept seeing the grinning face of a young man with a harelip. All the same, I drifted into sleep. The rocking movement of the carriage, the rhythm of the wheels on the rails, have always had a soothing effect on my nerves. I sleep well in trains and rocking chairs.

But not that night.

I woke to the sound of that familiar tapping; not at the door, but on the window glass not far from my head. The insistent tapping of someone who wanted to get in.

It was common enough for ticketless travellers to hang on to the carriage of a moving train, in the hope that someone would let them in. But they usually chose the crowded second or third-class compartments; a first-class traveller, often alone, was unlikely to let in a stranger who might well turn out to be a train robber.

I raised my head from my pillow, and there he was, clinging to the fast-moving train, his face pressed to the glass, his harelip revealing part of a broken tooth... I pulled down the shutters, blotting out his face. But, agile as a cat, he moved to the next window, the sneer still on his face. I pulled down that shutter too.

I pulled down all the shutters on his side of the carriage. He couldn't get in, bodily. But mentally, he was all over me.

Mind over matter. Well, I could apply my mind too. I shut my eyes and willed my tormentor to fall off the train!

No one fell off the train (at least no one was reported to have done so), but presently we slowed to a gradual stop and, when I pulled up the shutters of the window, I saw that we were at a station. Jalandhar, I think. The platform was brightly lit and there was no sign of Sohan. He must have jumped off the train as it slowed down. It was about one in the morning. A vendor brought me a welcome glass of hot tea, and life returned to normal.

◆

I did not see Sohan in the years that followed. Or rather, I saw many Sohans. For two or three years I was pursued by my 'familiar'. Wherever I went—and my work took me to different parts of the country—I found myself encountering young men with harelips and a menacing look. Pure imagination, of course. He had every reason to stay as far from me as possible.

Gradually, the 'sightings' died down. Young men with harelips became extremely rare. Perhaps they were all going in for corrective surgery.

The years passed, and I had forgotten my familiar. I had given up my job in Delhi and moved to the hills. I was a moderately successful writer, and a familiar figure on Mussoorie's Mall Road. Sometimes other writers came to see me, in my cottage under the deodars. One of them invited me to have dinner with him at the old Regal hotel, where he was staying. Before dinner, he took me to the bar for a drink.

'What will you have, whisky or vodka?'

No one seemed to drink anything else. I asked for some dark rum, and the barman went off in search of a bottle. When he returned and began pouring my drink, I noticed something slightly familiar about his features, his stance. He was almost bald, and he had a grey, drooping moustache that concealed most of his upper lip. He glanced at me and our eyes met. There was no sign of recognition. He smiled politely as he poured my drink. No, it definitely wasn't Sohan. He was too refined, for one thing. And he went about his duties without another glance in my direction.

Dinner over, I thanked my writer friend for his hospitality, and took the long walk home to my cottage. It was a dark, moonless night. No one followed me, no one came tapping on my bedroom window.

◆

Mussoorie had its charms. In my mind, every hill station is symbolized by a particular tree, even if it's not the dominant one. Dalhousie has its rhododendrons, Simla its deodars, Kasauli its pines, and Mussoorie its horse chestnuts. The monkeys would do their best to destroy the chestnuts, but I would collect those that were whole and plant them in people's gardens, whether they wanted them or not. The horse chestnut is a lovely tree to look at, even if you can't do anything with it!

My walks took me to the Regal from time to time, and occasionally I would relax in the bar, chatting to an old resident or a casual visitor, while the barman poured me a rum and soda. He never looked twice at me. And I never saw him outside that barroom. He appeared to be as much of a fixture as the moth-eaten antler-head on the wall, only he wasn't quite as moth-eaten.

'Efficient chap,' said Colonel Bhushan indicating the barman. 'And a great favourite with his mistress.'

'You mean the owner of this place?' I had only a vague idea of who owned what in the town. And in some cases the ownership was rather vague. But in the case of the Regal—Mrs Kapoor, a wealthy widow in her fifties, was very much in charge, all too visible an owner; well fleshed-out, ample-bosomed, with arms like rolling pins. Her staff trembled at her approach; but not, it seemed, the bartender, who led a charmed life, incapable of doing any wrong.

The lights went out, as they frequently do in this technological age, and the barman brought over our next round of drinks by candlelight.

By the light of a candle I caught a glimpse of the barman's features as he hovered over me. There was only the hint of a harelip, and the candle lit up his slanting eyes and prominent cheekbones. This was the only time I had a really close look at him.

◆

A week later I met Colonel Bhushan on the Mall. This was where all the gossip took place.

'Have you heard what happened last night at the Regal?' He wasted no time in getting to the news of the day.

A twinge of fear, of anticipation, ran through me. 'Nothing too terrible, I hope?'

'That barman chap—always thought he was a bit too smooth—stabbed the old lady, stabbed her two or three times, then plundered her room and made off with jewellery worth lakhs—as well as all the cash he could find!'

'How's the lady?'

'She'll survive. Tough old buffalo. But the rascal got away. By now he must be in Sirmur, or even across the Nepal border. Probably belongs to some criminal tribe.'

Yes, I thought, possibly a descendant of one of those robber gangs who harassed pilgrims on their way to the sacred shrines, or plundered traders from Tibet, or caravans to Samarkand... To rob and plunder still runs in the blood of the most harmless-looking people.

So the barman at the Regal was the same man I'd known in Dehradun and then encountered in Dalhousie. The passing of time had altered his features but not his way of life. By now he would probably be far from Mussoorie. But I had a feeling I'd see him again—if not here, then somewhere else. Each one of us had a 'familiar'—a presence we would rather do without—an unwelcome and menacing guest—and for me it is Sohan.

Where does he come from, where does he go? I doubt if I shall ever know.

But I have a feeling he'll turn up again one of these days. And then?

AMONG THE MAPLES AND OAKS

It isn't many years since I left Maplewood, but I wouldn't be surprised to hear that the cottage has disappeared. Already, during my last months there, the trees were being cut and the new road was being blasted out of the mountain. It would pass just below the old cottage. There were (as far as I know) no plans to blow up the house; but it was already shaky and full of cracks, and a few tremors, such as those produced by passing trucks, drilling machines and bulldozers, would soon bring the cottage to the ground.

If it has gone, don't write and tell me: I'd rather not know.

When I moved in, it had been nestling there among the oaks for over seventy years. It had become a part of the forest. Birds nestled in the eaves; beetles burrowed in the woodwork. Some denizens remained, even during my residence. And I was there—how long? Eight, nine years, I'm not sure; it was a timeless sort of place. Even the rent was paid only once a year, at a time of my choosing.

I first saw the cottage in late spring, when the surrounding forest was at its best—the oaks and maples in new leaf, the oak leaves a pale green, the maple leaves red and gold and bronze; this is the Himalayan maple, quite different from the North American maple; only the winged seed-pods are similar, twisting and turning in the breeze as they fall to the ground, so that the Garhwalis call it the Butterfly Tree.

There was one very tall, very old maple above the cottage, and this was probably the tree that gave the house its name. A portion of it was blackened where it had been struck by lightning, but the rest of it lived on; a favourite haunt of woodpeckers: the ancient peeling bark seemed to harbour any number of tiny insects, and the woodpeckers would be tapping away all day. A steep path ran down to the cottage. During heavy rain, it would become a watercourse and the earth would be washed away to leave it very stony and uneven. I first took this path to see Miss Mackenzie, an impoverished old lady who lived in two small rooms on the ground floor and who was acting on behalf of the owner. It was she who told me that the cottage was to-let provided she could remain in the portion downstairs.

Actually, the path ran straight across a landing and up to the front door of the first floor. It was the ground floor that was tucked away in the shadow of the hill; it was reached by a flight of steps, which also took the rush of water when the path was in flood.

Miss Mackenzie was eighty-six. I helped her up the steps and she opened the door for me. It led into an L-shaped room. There were two large windows, and when I pushed the first of these open, the forest seemed to rush upon me. From below, from the ravine, the deep-throated song of the whistling thrush burst upon me.

I told Miss Mackenzie I would take the place. She grew excited; it must have been lonely for her during the past several years, with most of the cottage lying empty, and only her old bearer and a mongrel dog for company. Her own house had been mortgaged to a moneylender. Her brothers and sisters were long dead.

I told her I would move in soon: my books were still in Delhi. She gave me the keys and I left a cheque with her. It was all done on an impulse—the decision to give up my job in Delhi, find a cheap house in a hill station, and return to freelance writing. It was a dream I'd had for some time; lack of money had made it difficult to realize. But then, I knew that if I was going to wait for money to come, I might have to wait until I was old and grey and full of sleep. I was thirty-five—still young enough to take a few risks. If the dream was to become reality, this was the time to do something about it.

I don't know what led me to Maplewood; it was the first place I saw, and I did not bother to see any others. The location was far from being ideal. It faced east, and stood in the shadow of the Balahissar Hill; so that while it received the early morning sun, it went without the evening sun.

There was no view of the snows and no view of the plains. In front stood Burnt Hill. But the forest below the cottage seemed full of possibilities, and the windows opening on to it probably decided the issue. In my romantic frame of my mind, I was susceptible to magic casements opening wide. I would make a window-seat and lie there on a summer's day, writing lyric poetry…

But long before that could happen I was opening tins of sardines and sharing them with Miss Mackenzie. And then Prem came along. And there were others, like Binya. I went away at times, but returned as soon as possible. Once you have lived with mountains, there is no escape. You belong to them.

GUESTS WHO FLY IN
FROM THE FOREST

When mist fills the Himalayan valleys, and heavy monsoon rain sweeps across the hills, it is natural for wild creatures to seek shelter. Any shelter is welcome in a storm—and sometimes my cottage in the forest is the most convenient refuge.

There is no doubt that I make things easier for all concerned by leaving most of my windows open—I am one of those peculiar people who like to have plenty of fresh air indoors—and if a few birds, beasts and insects come in too, they're welcome, provided they don't make too much of a nuisance of themselves.

I must confess that I did lose patience with a bamboo beetle who blundered in the other night and fell into the water jug. I rescued him and pushed him out of the window. A few seconds later he came whirring in again, and with unerring accuracy landed with a plop in the same jug. I fished him out once more and offered him the freedom of the night. But attracted no doubt by the light and warmth of my small sitting-room, he came buzzing back, circling the room like a helicopter looking for a good place to land. Quickly I covered the water jug. He landed in a bowl of wild dahlias, and I allowed him to remain there, comfortably curled up in the hollow of a flower.

Sometimes, during the day, a bird visits me—a deep purple whistling-thrush, hopping about on long dainty legs, peering to right and left, too nervous to sing. She perches on the windowsill, looking out at the rain. She does not permit any familiarity. But if I sit quietly in my chair, she will sit quietly on her windowsill, glancing quickly at me now and then just to make sure that I'm keeping my distance. When the rain stops, she glides away, and it is only then, confident in her freedom, that she bursts into full-throated song, her broken but haunting melody echoing down the ravine.

A squirrel comes sometimes, when his home in the oak tree gets waterlogged. Apparently he is a bachelor; anyway, he lives alone. He knows me well, this squirrel, and is bold enough to climb on to the dining-table looking for tidbits which he always finds, because I leave them there deliberately. Had I met him when he was a youngster, he would have learned to eat from my hand; but I have only been here a few months. I like it this way. I am not looking for pets: these are simply guests.

Last week, as I was sitting down at my desk to write a long-deferred article, I was startled to see an emerald-green praying mantis sitting on my writing pad. He peered up at me with his protruberant glass bead eyes, and I stared down at him through my reading glasses. When I gave him a prod, he moved off in a leisurely way. Later I found him examining the binding of Whitman's *Leaves of Grass*; perhaps he had found a succulent bookworm. He disappeared for a couple of days, and then I found him on the dressing-table, preening himself before the mirror. Perhaps I am doing him an injustice in assuming that he was preening. Maybe he thought he'd met another mantis and was simply trying to make contact. Anyway, he seemed fascinated by his reflection.

Out in the garden, I spotted another mantis, perched on the jasmine bush. Its arms were raised like a boxer's. Perhaps they're a pair, I thought, and went indoors and fetched my mantis and placed him on the jasmine bush, opposite his fellow insect. He did not like what he saw—no comparison with his own image!—and made off in a huff.

My most interesting visitor comes at night, when the lights are still burning—a tiny bat who prefers to fly in at the door, should it be open, and will use the window only if there's no alternative. His object in entering the house is to snap up the moths that cluster around the lamps.

All the bats I've seen fly fairly high, keeping near the ceiling as far as possible, and only descending to ear level (my ear level) when they must; but this particular bat flies in low, like a dive bomber, and does acrobatics amongst the furniture, zooming in and out of chair legs and under tables. Once, while careening about the room in this fashion, he passed straight between my legs.

Has his radar gone wrong, I wondered, or is he just plain crazy?

I went to my shelves of *Natural History* and looked up Bats, but could find no explanation for this erratic behaviour. As a last resort, I turned to an ancient volume, Sterndale's *Indian Mammalia* (Calcutta, 1884), and in it, to my delight, I found what I was looking for:

> ...a bat found near Mussoorie by Captain Hutton, on the southern range of hills at 5,500 feet; head and body, 1.4 inch; skims close to the ground, instead of flying high as bats generally do, Habitat, Jharipani, N.W. Himalayas.

Apparently the bat was rare even in 1884.

Perhaps I've come across one of the few surviving members of the species: Jharipani is only two miles from where I live. And I feel rather offended that modern authorities should have ignored this tiny bat; possibly they feel that it is already extinct. If so, I'm pleased to have rediscovered it. I am happy that it survives in my small corner of the woods, and I undertake to celebrate it in prose and verse.

GREAT SPIRITS OF THE TREES

Explore the history and mythology of almost any Indian tree, and you will find that at some point in our civilization it has held an important place in the minds and hearts of the people of this land.

During the rains, when the neem-pods fall and are crushed underfoot, they give out a strong refreshing aroma which lingers in the air for days. This is because the neem gives out more oxygen than most trees. When the ancient herbalists held that the neem was a great purifier of the air, and that its leaves, bark and sap had medicinal qualities, they were quite right, for the neem is still used in medicine today.

From the earliest times it was connected with the gods who protect us from disease. Some castes regarded the tree as sacred to Sitala, the smallpox goddess. When children fell ill, a branch of the neem was waved over them. The tree is said to have sprung from the nectar of the gods, and people still chew the leaves as a means of purification, both spiritual and physical. The tree is also connected with the sun, as in the story of Neembarak, 'The Sun in the Neem Tree'. The Sun God invited to dinner a man of the Bairagi tribe whose rules forbade him to eat except by daylight. Dinner was late, and as darkness fell, the Bairagi feared he would have to go hungry. But Suraj Narayan, the Sun God, descended from a neem tree and continued shining till dinner was over.

Why have so many trees been held sacred, not only in India but the world over?

To early man they were objects of awe and wonder. The mystery of their growth, the movement of their leaves and branches, the way they seemed to die and then come to life again in spring, the growth of the plant from the seed, all these happenings appeared as miracles—as indeed they are! And because of the wonderful growth of a tree, people began to suppose that it was occupied by spirits, and devotion to a tree became devotion to the spirit or tree god who occupied it. In *Puck of Pook's Hill,* Kipling wove some wonderful stories around Puck, the tree-spirit, and the sacred trees of Old England—oak, ash and thorn: 'I came into England with Oak, Ash, and Thorn, and when Oak, Ash and Thorn are gone, I shall go too.'

Among the Gonds of Central India, before a man cut a tree he had to beg its pardon for the injury he was about to inflict on it. He would not shake a tree at night because the tree spirit was asleep and might be disturbed. When a tree had to be felled, the Gonds would pour ghee on the stump, saying: 'Grow thou out of this, O Lord of the Forest, grow into a hundred shoots! May we grow with a thousand shoots.'

The beautiful mahua is a forest tree held sacred by a number of tribes. Early on the wedding morning, before he goes to fetch his bride, the Bagdi bridegroom goes through a mock marriage with a mahua tree. He embraces it and daubs it with vermilion, his right wrist is bound to it with thread, and after he is released from the tree the thread is used to attach a bunch of mahua leaves to his wrist.

There is a beautiful tradition connected with the sal tree. It is said that at the time of the Buddha's birth, his mother stretched out her hand to take hold of a branch of the sal and he was delivered. Sal trees are also said to have rendered homage to the Buddha at his death, letting fall on him their flowers out of season, and bending their branches to shade him.

Special respect is paid to trees growing near the graves of Muslim saints. Near the tomb of a famous saint, Musa Sohag, at

Ahmedabad, there used to be a large old champa tree—perhaps it is still there—the branches of which were hung with glass bangles. Those anxious to have children came and offered bangles to the saint—the number of bangles depending on the means of the supplicant. If the saint favoured a wish, the champa tree 'snatched up the bangles and wore them on its arms'.

Another spectacular tree which has its place in our folklore is the dhak, or palasa, which gave its name to the battlefield of Plassey. It has the habit of dropping its leaves when it flowers, the upper and outer branches standing out in sprays of scarlet and orange. The flowers are sometimes used to dye the powder scattered at Holi, the spring festival; and the wood, said to contain the seed of fire, is used in lighting the Holi bonfire. Legend tells us that the Sun God aimed an arrow at earth, and that it took root and became the palasa tree.

The babul (or keekar) is not very impressive to look at but it will grow almost anywhere in the plains, and there are a number of old beliefs associated with it. For instance, you can cure fever and headache at a babul tree if you tie seven cotton threads from your left big toe to your head, and from your head to a branch of the tree. Then you must embrace the trunk seven times. Try it sometime. You will be so busy tying threads that you will forget you ever had a headache! And there are no after-effects.

Another belief concerning the babul is that if you water it regularly for thirteen days, you acquire control over the spirit who occupies it. There is a story about a man in Saharanpur who did this, and when he died and his corpse was taken away for cremation, no sooner was his pyre lit than he got up and walked away!

In the folklore of India, the mango is the 'wish-fulfilling tree'. When you want to make a wish on a mango tree, shut your eyes and get someone to lead you to the tree; then rub mango blossoms in your hands, and make your wish. The favour granted lasts only for a year, and the charm must be performed again at the next flowering of the tree. In the spring, the young leaves and buds symbolize the darts of Manmatha, or Kamadeva, God of Love.

Another 'wishing tree', the Kalp-vriksha, is an enormous old mulberry that is still cared for at Joshimath in Garhwal. It is said to be the tree beneath which the great Sankaracharya often meditated during his sojourn in the Himalayas. Judging by its girth, it might well be over a thousand years old.

Whole forests have been held sacred, such as that in Berar which was dedicated to a particular temple; no one dared to buy or cut the trees. The sacred groves near Mathura, where Lord Krishna sported as a youth, were also protected for centuries. But now, alas, even the hallowed groves are disappearing, making way for the demands of an ever-increasing population. A pity, because every human needs a tree of his own. Even if you do not worship the tree spirit, you can love the tree.

WILD FLOWERS NEAR A MOUNTAIN STREAM

Below my house is a forest of oak and maple and Himalayan rhododendron. A path twists its way down through the trees, over an open ridge where red sorrel grows wild, and then steeply down through a tangle of thorn bushes, vines and rangal bamboo. At the bottom of the hill the path leads on to a grassy verge, surrounded by wild rose. A stream runs close by the verge, tumbling over smooth pebbles, over rocks worn yellow with age, on its way to the plains and the little Song River and finally to the sacred Ganges.

When I first discovered the stream it was April and the wild roses were flowering, small white blossoms lying in clusters. There were primroses on the hill slopes, and an occasional late-flowering rhododendron provided a splash of red against the dark green of the hill.

The St John's Wort was flowering profusely on small shrubs.

Many legends have grown around this flower of pure dazzling sunshine, which takes its family name—Hypericaceae—from the great Titan god Hyperion, who was the father of the Greek god of the sun, Apollo.

Is a friend of yours insane? Then get him to drink the sap from the leaves and stalks of the St John's Wort. He will be well again.

Are you hurt? If your wounds do not heal, take the juice and put it on the wound; and if the bleeding will not stop, take more juice.

Is your father bald? Then he must rise early one morning

and bathe his head with the dew from St John's Wort, and his hair will grow again.

Do you live on the Isle of Man? Then beware! Tread not on the St John's Wort after sunset, lest a fairy horseman arise and carry you off. He will land you anywhere.

These are all English or Irish superstitions, but the St John's Wort is as profuse in the lower ranges of the Himalayas as it is anywhere in Europe.

A spotted forktail, a bird of the Himalayan streams, was much in evidence during those early visits. It moved nimbly over the boulders with a fairy tread, and continually wagged its tail.

In May and June, when the hills are always brown and dry, it remained cool and green near the stream, where ferns and maidenhair and long grasses continued to thrive. Downstream I found a cave with water dripping from the roof, the water spangled gold and silver in the shafts of sunlight that pushed through the slits in the cave roof. Few people came there. Sometimes a milkman or a coal-burner would cross the stream on his way to a village; but the nearby hill station's summer visitors had not discovered this haven of wild and green things.

The monkeys—langurs, with white and silver-grey fur, black faces and long swishing tails—had discovered the place, but they kept to the trees and sunlit slopes. They grew quite accustomed to my presence, and carried on with their work and play as though I did not exist. The young ones scuffled and wrestled like boys, while their parents attended to each other's toilets, stretching themselves out on the grass, beautiful animals with slim waists and long sinewy legs, and tails full of character. They were clean and polite, much nicer than the red monkeys of the plains.

During the rains the stream became a rushing torrent, bushes and small trees were swept away, and the friendly murmur of the water became a threatening boom. I did not visit the spot very often. There were leeches in the long grass, and they would fasten themselves onto my legs and feast on my blood. But it was always worthwhile tramping through the forest to feast my eyes on the foliage that sprang up in tropical profusion—soft, spongy moss; great stag ferns on the trunks of trees; mysterious and sometimes evil-looking orchids; the climbing convolvulus opening its purple secrets to the morning sun; and the wood sorrel, or oxalis—so named because of the oxalic acid derived from its roots—with its clover-like leaflets, which fold down like umbrellas at the first sign of rain.

And then, after a November hailstorm, it was winter, and one could not lie on the frostbitten grass. The sound of the stream was the same, but I missed the birds.

It snowed—the snow lay heavy on the branches of the oak trees and piled up in the culverts—and the grass and the ferns and wild flowers were pressed to sleep beneath a cold white blanket; but the stream flowed on, pushing its way through and under the whiteness, towards another river, towards another spring.

THE SCHOOL AMONG THE PINES

1

A leopard, lithe and sinewy, drank at the mountain stream, and then lay down on the grass to bask in the late February sunshine. Its tail twitched occasionally and the animal appeared to be sleeping. At the sound of distant voices it raised its head to listen, then stood up and leapt lightly over the boulders in the stream, disappearing among the trees on the opposite bank.

A minute or two later, three children came walking down the forest path. They were a girl and two boys, and they were singing in their local dialect an old song they had learnt from their grandparents.

Five more miles to go!
We climb through rain and snow.
A river to cross...
A mountain to pass...
Now we've four more miles to go!

Their school satchels looked new, their clothes had been washed and pressed. Their loud and cheerful singing startled a Spotted Forktail. The bird left its favourite rock in the stream and flew down the dark ravine.

'Well, we have only three more miles to go,' said the bigger boy, Prakash, who had been this way hundreds of times. 'But first we have to cross the stream.'

He was a sturdy twelve-year-old with eyes like raspberries and a mop of bushy hair that refused to settle down on his head. The girl and her small brother were taking this path for the first time.

'I'm feeling tired, Bina,' said the little boy.

Bina smiled at him, and Prakash said, 'Don't worry, Sonu, you'll get used to the walk. There's plenty of time.' He glanced at the old watch he'd been given by his grandfather. It needed constant winding. 'We can rest here for five or six minutes.'

They sat down on a smooth boulder and watched the clear water of the shallow stream tumbling downhill. Bina examined the old watch on Prakash's wrist. The glass was badly scratched and she could barely make out the figures on the dial. 'Are you sure it still gives the right time?' she asked.

'Well, it loses five minutes every day, so I put it ten minutes forward at night. That means by morning it's quite accurate! Even our teacher, Mr Mani, asks me for the time. If he doesn't ask, I tell him! The clock in our classroom keeps stopping.'

They removed their shoes and let the cold mountain water run over their feet. Bina was the same age as Prakash. She had pink cheeks, soft brown eyes, and hair that was just beginning to lose its natural curls. Hers was a gentle face, but a determined little chin showed that she could be a strong person. Sonu, her younger brother, was ten. He was a thin boy who had been sickly as a child but was now beginning to fill out. Although he did not look very athletic, he could run like the wind.

Bina had been going to school in her own village of Koli, on the other side of the mountain. But it had been a primary school, finishing at Class Five. Now, in order to study in the Sixth, she would have to walk several miles every day to Nauti, where there was a high school going up to the Eighth. It had been decided that Sonu would also shift to the new school, to give Bina company. Prakash, their neighbour in Koli, was already a pupil at the Nauti school. His mischievous nature, which sometimes got him into trouble, had resulted in his having to repeat a year.

But this didn't seem to bother him. 'What's the hurry?' he had told his indignant parents. 'You're not sending me to a foreign land when I finish school. And our cows aren't running away, are they?'

'You would prefer to look after the cows, wouldn't you?' asked Bina, as they got up to continue their walk.

'Oh, school's all right. Wait till you see old Mr Mani. He always gets our names mixed up, as well as the subjects he's supposed to be teaching. At our last lesson, instead of maths, he gave us a geography lesson!'

'More fun than maths,' said Bina.

'Yes, but there's a new teacher this year. She's very young, they say, just out of college. I wonder what she'll be like.'

Bina walked faster and Sonu had some trouble keeping up with them. She was excited about the new school and the prospect of different surroundings. She had seldom been outside her own village, with its small school and single ration shop. The day's routine never varied—helping her mother in the fields or with household tasks like fetching water from the spring or cutting grass and fodder for the cattle. Her father, who was a soldier, was away for nine months in the year and Sonu was still too small for the heavier tasks.

As they neared Nauti village, they were joined by other children coming from different directions. Even where there were no major roads, the mountains were full of little lanes and short cuts. Like a game of snakes and ladders, these narrow paths zigzagged around the hills and villages, cutting through fields and crossing narrow ravines until they came together to form a fairly busy road along which mules, cattle and goats joined the throng.

Nauti was a fairly large village, and from here a broader but dustier road started for Tehri. There was a small bus, several trucks and (for part of the way) a road-roller. The road hadn't been completed because the heavy diesel roller couldn't take the steep climb to Nauti. It stood on the roadside halfway up the road from Tehri.

Prakash knew almost everyone in the area, and exchanged greetings and gossip with other children as well as with muleteers, bus drivers, milkmen and labourers working on the road. He loved telling everyone the time, even if they weren't interested.

'It's nine o'clock,' he would announce, glancing at his wrist. 'Isn't your bus leaving today?'

'Off with you!' the bus driver would respond, 'I'll leave when I'm ready.'

As the children approached Nauti, the small flat school buildings came into view on the outskirts of the village, fringed with a line of long-leaved pines. A small crowd had assembled on the playing field. Something unusual seemed to have happened. Prakash ran forward to see what it was all about. Bina and Sonu stood aside, waiting in a patch of sunlight near the boundary wall.

Prakash soon came running back to them. He was bubbling over with excitement.

'It's Mr Mani!' he gasped. 'He's disappeared! People are saying a leopard must have carried him off!'

2

Mr Mani wasn't really old. He was about fifty-five and was expected to retire soon. But for the children, adults over forty seemed ancient! And Mr Mani had always been a bit absent-minded, even as a young man.

He had gone out for his early morning walk, saying he'd be back by eight o'clock, in time to have his breakfast and be ready for class. He wasn't married, but his sister and her husband stayed with him. When it was past nine o'clock his sister presumed he'd stopped at a neighbour's house for breakfast (he loved tucking into other people's breakfast) and that he had gone on to school from there. But when the school bell rang at ten o'clock, and everyone but Mr Mani was present, questions were asked and guesses were made.

No one had seen him return from his walk and enquiries made in the village showed that he had not stopped at anyone's house. For Mr Mani to disappear was puzzling; for him to disappear without his breakfast was extraordinary.

Then a milkman returning from the next village said he had seen a leopard sitting on a rock on the outskirts of the pine forest. There had been talk of a cattle-killer in the valley, of leopards and other animals being displaced by the construction of a dam. But as yet no one had heard of a leopard attacking a man. Could Mr Mani have been its first victim? Someone found a strip of red cloth entangled in a blackberry bush and went running through the village showing it to everyone. Mr Mani had been known to wear red pyjamas. Surely, he had been seized and eaten! But where were his remains? And why had he been in his pyjamas?

Meanwhile, Bina and Sonu and the rest of the children had followed their teachers into the school playground. Feeling a little lost, Bina looked around for Prakash. She found herself facing a dark slender young woman wearing spectacles, who must have been in her early twenties—just a little too old to be another student. She had a kind expressive face and she seemed a little concerned by all that had been happening.

Bina noticed that she had lovely hands; it was obvious that the new teacher hadn't milked cows or worked in the fields!

'You must be new here,' said the teacher, smiling at Bina. 'And is this your little brother?'

'Yes, we've come from Koli village. We were at school there.'

'It's a long walk from Koli. You didn't see any leopards, did you? Well, I'm new too. Are you in the Sixth class?'

'Sonu is in the Third. I'm in the Sixth.'

'Then I'm your new teacher. My name is Tania Ramola. Come along, let's see if we can settle down in our classroom.'

Mr Mani turned up at twelve o'clock, wondering what all the fuss was about. No, he snapped, he had not been attacked by a leopard; and yes, he had lost his pyjamas and would someone kindly return them to him?

'How did you lose your pyjamas, sir?' asked Prakash.

'They were blown off the washing line!' snapped Mr Mani.

After much questioning, Mr Mani admitted that he had gone further than he had intended, and that he had lost his way coming back. He had been a bit upset because the new teacher, a slip of a girl, had been given charge of the Sixth, while he was still with the Fifth, along with that troublesome boy Prakash, who kept on reminding him of the time! The headmaster had explained that as Mr Mani was due to retire at the end of the year, the school did not wish to burden him with a senior class. But Mr Mani looked upon the whole thing as a plot to get rid of him. He glowered at Miss Ramola whenever he passed her. And when she smiled back at him, he looked the other way!

Mr Mani had been getting even more absent-minded of late—putting on his shoes without his socks, wearing his homespun waistcoat inside out, mixing up people's names and, of course, eating other people's lunches and dinners. His sister had made a special mutton broth (*pai*) for the postmaster, who was down with flu and had asked Mr Mani to take it over in a thermos. When the postmaster opened the thermos, he found only a few drops of broth at the bottom—Mr Mani had drunk the rest somewhere along the way.

When sometimes Mr Mani spoke of his coming retirement, it was to describe his plans for the small field he owned just behind the house. Right now, it was full of potatoes, which did not require much looking after; but he had plans for growing dahlias, roses, French beans, and other fruits and flowers.

The next time he visited Tehri, he promised himself, he would buy some dahlia bulbs and rose cuttings. The monsoon season would be a good time to put them down. And meanwhile, his potatoes were still flourishing.

3

Bina enjoyed her first day at the new school. She felt at ease with Miss Ramola, as did most of the boys and girls in her class. Tania Ramola had been to distant towns such as Delhi and Lucknow—places they had only read about—and it was said that she had a brother who was a pilot and flew planes all over the world. Perhaps he'd fly over Nauti some day!

Most of the children had, of course, seen planes flying overhead, but none of them had seen a ship, and only a few had been on a train. Tehri mountain was far from the railway and hundreds of miles from the sea. But they all knew about the big dam that was being built at Tehri, just forty miles away.

Bina, Sonu and Prakash had company for part of the way home, but gradually the other children went off in different directions. Once they had crossed the stream, they were on their own again.

It was a steep climb all the way back to their village. Prakash had a supply of peanuts which he shared with Bina and Sonu, and at a small spring they quenched their thirst.

When they were less than a mile from home, they met a postman who had finished his round of the villages in the area and was now returning to Nauti.

'Don't waste time along the way,' he told them. 'Try to get home before dark.'

'What's the hurry?' asked Prakash, glancing at his watch. 'It's only five o'clock.'

'There's a leopard around. I saw it this morning, not far from the stream. No one is sure how it got here. So don't take any chances. Get home early.'

'So there really is a leopard,' said Sonu.

They took his advice and walked faster, and Sonu forgot to complain about his aching feet.

They were home well before sunset.

There was a smell of cooking in the air and they were hungry.

'Cabbage and roti,' said Prakash gloomily. 'But I could eat anything today.' He stopped outside his small slate-roofed house, and Bina and Sonu waved him goodbye, then carried on across a couple of ploughed fields until they reached their small stone house.

'Stuffed tomatoes,' said Sonu, sniffing just outside the front door.

'And lemon pickle,' said Bina, who had helped cut, sun and salt the lemons a month previously.

Their mother was lighting the kitchen stove. They greeted her with great hugs and demands for an immediate dinner. She was a good cook who could make even the simplest of dishes taste delicious. Her favourite saying was, 'Homemade pai is better than chicken soup in Delhi,' and Bina and Sonu had to agree.

Electricity had yet to reach their village, and they took their meal by the light of a kerosene lamp. After the meal, Sonu settled down to do a little homework, while Bina stepped outside to look at the stars.

Across the fields, someone was playing a flute. *It must be Prakash*, thought Bina. *He always breaks off on the high notes.*

But the flute music was simple and appealing, and she began singing softly to herself in the dark.

4

Mr Mani was having trouble with the porcupines. They had been getting into his garden at night and digging up and eating his potatoes. From his bedroom window—left open, now that the mild-April weather had arrived—he could listen to them enjoying the vegetables he had worked hard to grow. Scrunch, scrunch! *Katar, katar*, as their sharp teeth sliced through the largest and juiciest of potatoes. For Mr Mani it was as though they were biting through his own flesh. And the sound of them digging industriously as they rooted up those healthy, leafy plants made him tremble with rage and indignation. The unfairness of it all!

Yes, Mr Mani hated porcupines. He prayed for their destruction, their removal from the face of the earth. But, as his friends were quick to point out, 'Bhagwan protected porcupines too,' and in any case you could never see the creatures or catch them; they were completely nocturnal.

Mr Mani got out of bed every night, torch in one hand, a stout stick in the other, but as soon as he stepped into the garden the crunching and digging stopped and he was greeted by the most infuriating of silences. He would grope around in the dark, swinging wildly with the stick, but not a single porcupine was to be seen or heard. As soon as he was back in bed—the sounds would start all over again. Scrunch, scrunch, *katar, katar*...

Mr Mani came to his class tired and dishevelled, with rings beneath his eyes and a permanent frown on his face. It took some time for his pupils to discover the reason for his misery, but when they did, they felt sorry for their teacher and took to discussing ways and means of saving his potatoes from the porcupines.

It was Prakash who came up with the idea of a moat or waterditch. 'Porcupines don't like water,' he said knowledgeably.

'How do you know?' asked one of his friends.

'Throw water on one and see how it runs! They don't like getting their quills wet.'

There was no one who could disprove Prakash's theory, and the class fell in with the idea of building a moat, especially as it meant getting most of the day off.

'Anything to make Mr Mani happy,' said the headmaster, and the rest of the school watched with envy as the pupils of Class Five, armed with spades and shovels collected from all parts of the village, took up their positions around Mr Mani's potato field and began digging a ditch.

By evening the moat was ready, but it was still dry and the porcupines got in again that night and had a great feast.

'At this rate,' said Mr Mani gloomily, 'there won't be any potatoes left to save.'

But next day Prakash and the other boys and girls managed to divert the water from a stream that flowed past the village. They had the satisfaction of watching it flow gently into the ditch. Everyone went home in a good mood. By nightfall, the ditch had overflowed, the potato field was flooded, and Mr Mani found himself trapped inside his house. But Prakash and his friends had won the day. The porcupines stayed away that night!

A month had passed, and wild violets, daisies and buttercups now sprinkled the hill slopes, and on her way to school Bina gathered enough to make a little posy. The bunch of flowers fitted easily into an old inkwell. Miss Ramola was delighted to find this little display in the middle of her desk.

'Who put these here?' she asked in surprise.

Bina kept quiet, and the rest of the class smiled secretively. After that, they took turns bringing flowers for the classroom.

On her long walks to school and home again, Bina became aware that April was the month of new leaves. The oak leaves were bright green above and silver beneath, and when they rippled in the breeze they were like clouds of silvery green. The path was strewn with old leaves, dry and crackly. Sonu loved kicking them around.

Clouds of white butterflies floated across the stream. Sonu was chasing a butterfly when he stumbled over something dark and repulsive. He went sprawling on the grass. When he got to his feet, he looked down at the remains of a small animal.

'Bina! Prakash! Come quickly!' he shouted.

It was part of a sheep, killed some days earlier by a much larger animal.

'Only a leopard could have done this,' said Prakash.

'Let's get away, then,' said Sonu. 'It might still be around!'

'No, there's nothing left to eat. The leopard will be hunting elsewhere by now. Perhaps it's moved on to the next valley.'

'Still, I'm frightened,' said Sonu. 'There may be more leopards!'

Bina took him by the hand. 'Leopards don't attack humans!' she said.

'They will, if they get a taste for people!' insisted Prakash.

'Well, this one hasn't attacked any people as yet,' said Bina, although she couldn't be sure. Hadn't there been rumours of a leopard attacking some workers near the dam? But she did not want Sonu to feel afraid, so she did not mention the story. All she said was, 'It has probably come here because of all the activity near the dam.'

All the same, they hurried home. And for a few days, whenever they reached the stream, they crossed over very quickly, unwilling to linger too long at that lovely spot.

5

A few days later, a school party was on its way to Tehri to see the new dam that was being built.

Miss Ramola had arranged to take her class, and Mr Mani, not wishing to be left out, insisted on taking his class as well. That meant there were about fifty boys and girls taking part in the outing. The little bus could only take thirty. A friendly truck driver agreed to take some children if they were prepared to sit on sacks of potatoes. And Prakash persuaded the owner of the diesel roller to turn it round and head it back to Tehri—with him and a couple of friends up on the driving seat.

Prakash's small group set off at sunrise, as they had to walk some distance in order to reach the stranded road roller. The bus left at 9.00 a.m. with Miss Ramola and her class, and Mr Mani and some of his pupils. The truck was to follow later.

It was Bina's first visit to a large town and her first bus ride.

The sharp curves along the winding, downhill road made several children feel sick. The bus driver seemed to be in a tearing hurry. He took them along at rolling, rollicking speed, which made Bina feel quite giddy. She rested her head on her arms and refused to look out of the window. Hairpin bends and cliff edges, pine forests and snowcapped peaks, all swept past her, but she felt too ill to want to look at anything. It was just as well—those sudden drops, hundreds of feet to the valley below, were quite frightening. Bina began to wish that she hadn't come—or that she had joined Prakash on the road roller instead!

Miss Ramola and Mr Mani didn't seem to notice the lurching and groaning of the old bus. They had made this journey many times. They were busy arguing about the advantages and disadvantages of large dams—an argument that was to continue on and off for much of the day; sometimes in Hindi, sometimes in English, sometimes in the local dialect!

Meanwhile, Prakash and his friends had reached the roller. The driver hadn't turned up, but they managed to reverse it and get it going in the direction of Tehri. They were soon overtaken by both the bus and the truck but kept moving along at a steady chug. Prakash spotted Bina at the window of the bus and waved cheerfully. She responded feebly.

Bina felt better when the road levelled out near Tehri. As they crossed an old bridge over the wide river, they were startled by a loud bang that made the bus shudder. A cloud of dust rose above the town.

'They're blasting the mountain,' said Miss Ramola.

'End of a mountain,' said Mr Mani mournfully.

While they were drinking cups of tea at the bus stop, waiting for the potato truck and the road roller, Miss Ramola and Mr Mani continued their argument about the dam. Miss Ramola maintained that it would bring electric power and water for irrigation to large areas of the country, including the surrounding area. Mr Mani declared that it was a menace, as it was situated in an earthquake zone. There would be a terrible disaster if the dam burst! Bina found it all very confusing. *And what about the animals in the area*, she wondered. *What would happen to them?*

The argument was becoming quite heated when the potato truck arrived. There was no sign of the road roller, so it was decided that Mr Mani should wait for Prakash and his friends while Miss Ramola's group went ahead.

Some eight or nine miles before Tehri the road roller had broken down, and Prakash and his friends were forced to walk. They had not gone far, however, when a mule train came along—five or six mules that had been delivering sacks of grain in Nauti. A boy rode on the first mule, but the others had no loads.

'Can you give us a ride to Tehri?' called Prakash.

'Make yourselves comfortable,' said the boy.

There were no saddles, only gunny sacks strapped on to the mules with rope. They had a rough but jolly ride down to the Tehri bus stop. None of them had ever ridden mules; but they had saved at least an hour on the road.

Looking around the bus stop for the rest of the party, they could find no one from their school. And Mr Mani, who should have been waiting for them, had vanished.

6

Tania Ramola and her group had taken the steep road to the hill above Tehri. Half an hour's climbing brought them to a little plateau which overlooked the town, the river and the dam site.

The earthworks for the dam were only just coming up, but a wide tunnel had been bored through the mountain to divert the river into another channel. Down below, the old town was still spread out across the valley and from a distance it looked quite charming and picturesque.

'Will the whole town be swallowed up by the waters of the dam?' asked Bina.

'Yes, all of it,' said Miss Ramola. 'The clock tower and the old palace. The long bazaar, and the temples, the schools and the jail, and hundreds of houses, for many miles up the valley. All those people will have to go—thousands of them! Of course, they'll be resettled elsewhere.'

'But the town's been here for hundreds of years,' said Bina. 'They were quite happy without the dam, weren't they?'

'I suppose they were. But the dam isn't just for them—it's for the millions who live further downstream, across the plains.'

'And it doesn't matter what happens to this place?'

'The local people will be given new homes, somewhere else.'

Miss Ramola found herself on the defensive and decided to change the subject. 'Everyone must be hungry. It's time we had our lunch.'

Bina kept quiet. She didn't think the local people would want to go away. And it was a good thing, she mused, that there was only a small stream and not a big river running past her village. To be uprooted like this—a town and hundreds of villages—and put down somewhere on the hot, dusty plains—seemed to her unbearable.

'Well, I'm glad I don't live in Tehri,' she said.

She did not know it, but all the animals and most of the birds had already left the area. The leopard had been among them.

They walked through the colourful, crowded bazaar, where fruit sellers did business beside silversmiths, and pavement vendors sold everything from umbrellas to glass bangles. Sparrows attacked sacks of grain, monkeys made off with bananas, and stray cows and dogs rummaged in refuse bins, but nobody took any notice. Music blared from radios. Buses blew their horns. Sonu bought a whistle to add to the general din, but Miss Ramola told him to put it away. Bina had kept ten rupees aside, and now she used it to buy a cotton head scarf for her mother.

As they were about to enter a small restaurant for a meal, they were joined by Prakash and his companions; but of Mr Mani there was still no sign.

'He must have met one of his relatives,' said Prakash. 'He has relatives everywhere.'

After a simple meal of rice and lentils, they walked the length of the bazaar without seeing Mr Mani. At last, when they were about to give up the search, they saw him emerge from a bylane, a large sack slung over his shoulder.

'Sir, where have you been?' asked Prakash. 'We have been looking for you everywhere.'

On Mr Mani's face was a look of triumph.
'Help me with this bag,' he said breathlessly.
'You've bought more potatoes, sir,' said Prakash.
'Not potatoes, boy. Dahlia bulbs!'

7

It was dark by the time they were all back in Nauti. Mr Mani had refused to be separated from his sack of dahlia bulbs, and had been forced to sit in the back of the truck with Prakash and most of the boys.

Bina did not feel so ill on the return journey. Going uphill was definitely better than going downhill! But by the time the bus reached Nauti it was too late for most of the children to walk back to the more distant villages. The boys were put up in different homes, while the girls were given beds in the school verandah.

The night was warm and still. Large moths fluttered around the single bulb that lit the verandah. Counting moths, Sonu soon fell asleep. But Bina stayed awake for some time, listening to the sounds of the night. A nightjar went *tonk-tonk* in the bushes, and somewhere in the forest an owl hooted softly. The sharp call of a barking deer travelled up the valley, from the direction of the stream. Jackals kept howling. It seemed that there were more of them than ever before.

Bina was not the only one to hear the barking deer. The leopard, stretched full length on a rocky ledge, heard it too. The leopard raised its head and then got up slowly. The deer was its natural prey. But there weren't many left, and that was why the leopard, robbed of its forest by the dam, had taken to attacking dogs and cattle near the villages.

As the cry of the barking deer sounded nearer, the leopard left its lookout point and moved swiftly through the shadows towards the stream.

8

In early June the hills were dry and dusty, and forest fires broke out, destroying shrubs and trees, killing birds and small animals. The resin in the pines made these trees burn more fiercely, and the wind would take sparks from the trees and carry them into the dry grass and leaves, so that new fires would spring up before the old ones had died out. Fortunately, Bina's village was not in the pine belt; the fires did not reach it. But Nauti was surrounded by a fire that raged for three days, and the children had to stay away from school.

And then, towards the end of June, the monsoon rains arrived and there was an end to forest fires. The monsoon lasts three months and the lower Himalayas would be drenched in rain, mist and cloud for the next three months.

The first rain arrived while Bina, Prakash and Sonu were returning home from school. Those first few drops on the dusty path made them cry out with excitement. Then the rain grew heavier and a wonderful aroma rose from the earth.

'The best smell in the world!' exclaimed Bina.

Everything suddenly came to life. The grass, the crops, the trees, the birds. Even the leaves of the trees glistened and looked new.

That first wet weekend, Bina and Sonu helped their mother plant beans, maize and cucumbers. Sometimes, when the rain was very heavy, they had to run indoors. Otherwise they worked in the rain, the soft mud clinging to their bare legs.

Prakash now owned a black dog with one ear up and one ear down. The dog ran around getting in everyone's way, barking at cows, goats, hens and humans, without frightening any of them. Prakash said it was a very clever dog, but no one else seemed to think so. Prakash also said it would protect the village from the leopard, but others said the dog would be the first to be taken—he'd run straight into the jaws of Mr Spots!

In Nauti, Tania Ramola was trying to find a dry spot in the quarters she'd been given. It was an old building and the roof was leaking in several places. Mugs and buckets were scattered about the floor in order to catch the drip.

Mr Mani had dug up all his potatoes and presented them to the friends and neighbours who had given him lunches and dinners. He was having the time of his life, planting dahlia bulbs all over his garden.

'I'll have a field of many-coloured dahlias!' he announced. 'Just wait till the end of August!'

'Watch out for those porcupines,' warned his sister. 'They eat dahlia bulbs too!'

Mr Mani made an inspection tour of his moat, no longer in flood, and found everything in good order. Prakash had done his job well.

Now, when the children crossed the stream, they found that the water level had risen by about a foot. Small cascades had turned into waterfalls. Ferns had sprung up on the banks. Frogs chanted.

Prakash and his dog dashed across the stream. Bina and Sonu followed more cautiously. The current was much stronger now and the water was almost up to their knees. Once they had crossed the stream, they hurried along the path, anxious not to be caught in a sudden downpour.

By the time they reached school, each of them had two or three leeches clinging to their legs. They had to use salt to remove them. The leeches were the most troublesome part of the rainy season. Even the leopard did not like them. It could not lie in the long grass without getting leeches on its paws and face.

One day, when Bina, Prakash and Sonu were about to cross

the stream they heard a low rumble, which grew louder every second. Looking up at the opposite hill, they saw several trees shudder, tilt outwards and begin to fall. Earth and rocks bulged out from the mountain, then came crashing down into the ravine.

'Landslide!' shouted Sonu.

'It's carried away the path,' said Bina. 'Don't go any further.'

There was a tremendous roar as more rocks, trees and bushes fell away and crashed down the hillside.

Prakash's dog, who had gone ahead, came running back, tail between his legs.

They remained rooted to the spot until the rocks had stopped falling and the dust had settled. Birds circled the area, calling wildly. A frightened barking deer ran past them.

'We can't go to school now,' said Prakash. 'There's no way around.'

They turned and trudged home through the gathering mist.

In Koli, Prakash's parents had heard the roar of the landslide. They were setting out in search of the children when they saw them emerge from the mist, waving cheerfully.

9

They had to miss school for another three days, and Bina was afraid they might not be able to take their final exams. Although Prakash was not really troubled at the thought of missing exams, he did not like feeling helpless just because their path had been swept away. So he explored the hillside until he found a goat track going around the mountain. It joined up with another path near Nauti. This made their walk longer by a mile, but Bina did not mind. It was much cooler now that the rains were in full swing.

The only trouble with the new route was that it passed close to the leopard's lair. The animal had made this area its own since being forced to leave the dam area.

One day Prakash's dog ran ahead of them, barking furiously. Then he ran back, whimpering.

'He's always running away from something,' observed Sonu. But a minute later he understood the reason for the dog's fear.

They rounded a bend and Sonu saw the leopard standing in their way. They were struck dumb—too terrified to run. It was a strong, sinewy creature. A low growl rose from its throat. It seemed ready to spring.

They stood perfectly still, afraid to move or say a word. And the leopard must have been equally surprised. It stared at them for a few seconds, then bounded across the path and into the oak forest.

Sonu was shaking. Bina could hear her heart hammering. Prakash could only stammer: 'Did you see the way he sprang? Wasn't he beautiful?'

He forgot to look at his watch for the rest of the day.

A few days later Sonu stopped and pointed to a large outcrop of rock on the next hill.

The leopard stood far above them, outlined against the sky. It looked strong, majestic. Standing beside it were two young cubs.

'Look at those little ones!' exclaimed Sonu.

'So it's a female, not a male,' said Prakash.

'That's why she was killing so often,' said Bina. 'She had to feed her cubs too.'

They remained still for several minutes, gazing up at the leopard and her cubs. The leopard family took no notice of them.

'She knows we are here,' said Prakash, 'but she doesn't care. She knows we won't harm them.'

'We are cubs too!' said Sonu.

'Yes,' said Bina. 'And there's still plenty of space for all of us. Even when the dam is ready there will still be room for leopards and humans.'

10

The school exams were over. The rains were nearly over too. The landslide had been cleared, and Bina, Prakash and Sonu were once again crossing the stream.

There was a chill in the air, for it was the end of September.

Prakash had learnt to play the flute quite well, and he played on the way to school and then again on the way home. As a result he did not look at his watch so often.

One morning they found a small crowd in front of Mr Mani's house.

'What could have happened?' wondered Bina. 'I hope he hasn't got lost again.'

'Maybe he's sick,' said Sonu.

'Maybe it's the porcupines,' said Prakash.

But it was none of these things.

Mr Mani's first dahlia was in bloom, and half the village had turned out to look at it! It was a huge red double dahlia, so heavy that it had to be supported with sticks. No one had ever seen such a magnificent flower!

Mr Mani was a happy man. And his mood only improved over the coming week, as more and more dahlias flowered—crimson, yellow, purple, mauve, white—button dahlias, pompom dahlias, spotted dahlias, striped dahlias... Mr Mani had them all! A dahlia even turned up on Tania Romola's desk—he got on quite well with her now—and another brightened up the headmaster's study.

A week later, on their way home—it was almost the last day of the school term—Bina, Prakash and Sonu talked about what they might do when they grew up.

'I think I'll become a teacher,' said Bina. 'I'll teach children about animals and birds, and trees and flowers.'

'Better than maths!' said Prakash.

'I'll be a pilot,' said Sonu. 'I want to fly a plane like Miss Ramola's brother.'

'And what about you, Prakash?' asked Bina.

Prakash just smiled and said, 'Maybe I'll be a flute player,' and he put the flute to his lips and played a sweet melody.

'Well, the world needs flute players too,' said Bina, as they fell into step beside him.

The leopard had been stalking a barking deer. She paused when she heard the flute and the voices of the children. Her own young ones were growing quickly, but the girl and the two boys did not look much older.

They had started singing their favourite song again:

Five more miles to go!
We climb through rain and snow.
A river to cross...
A mountain to pass...
Now we've four more miles to go!

The leopard waited until they had passed, before returning to the trail of the barking deer.

IN THE GARDEN OF MY DREAMS

The cosmos has all the genius of simplicity. The plant stands tall and erect; its foliage is uncomplicated, its inflorescence are bold, fresh, cheerful. Any flower, from a rose to a rhododendron, can be complicated. The cosmos is splendidly simple.

No wonder it takes its name from the Greek 'cosmos', meaning the universe as an ordered whole—the sum total of experience! For this unpretentious flower does seem to sum it all up: perfection without apparent striving for it, the artistry of the South American footballer! Needless to say, it came from tropical America.

And growing it is no trouble. A handful of seeds thrown in a waste patch or on a grassy hill slope, and a few months later there they are, en masse, doing their samba in the sunshine. They are almost wild, but not quite. They need very little attention, but if you take them too much for granted they will go away the following year. Simple they may be, but not insensitive. They need plenty of space. And as my own small apartment cannot accommodate them, they definitely belong to my dream garden.

My respect for the cosmos goes back to my childhood when I wandered into what seemed like a forest of these flowers, all twice my height (I must have been five at the time) but looking down on me in the friendliest way, their fine feathery foliage giving off a faint aroma. Now when I find them flowering on the Himalayan hillsides in mellow October sunshine, they are like old friends and I greet them accordingly, pressing my face to their petals.

Not everyone likes the cosmos. I have met some upper-class ladies (golf club members) who complain that it gives them hay fever, and they use this as an excuse to root out all cosmos from their gardens. I expect they are just being snobbish. There are other flowers that give off just as much pollen dust.

I have noticed the same snobbishness in regard to marigolds, especially the smaller Indian variety. 'Cultivated' people won't cultivate these humble but attractive flowers. Is it because they are used for making garlands? Or because they are not delicately scented? Or because they are so easily grown in the backyards of homes?

My grandparents once went to war with each other over the marigold. Grandfather had grown a few in one corner of the garden. Just as they began flowering, they vanished—Granny had removed them overnight! There was a row, and my grandparents did not speak to each other for several days. Then, by calling them 'French' marigolds, Grandfather managed to reintroduce them to the garden. Granny liked the idea of having something 'French' in her garden. Such is human nature!

Sometimes a wildflower can put its more spectacular garden cousins to shame. I am thinking now of the Commelina, which I discover in secret places after the rains have passed. Its bright sky-blue flowers take my breath away. It has a sort of unguarded innocence that is beyond corruption.

Wild roses give me more pleasure than the sophisticated domestic variety. On a walk in the Himalayan foothills I have encountered a number of these shrubs and climbers—the ineptly named Dog Rose, sparkling white in summer; the Sweet Briar with its deep pink petals and bright red rosehips; the Trailing Rose, found in shady places; and the wild Raspberry (the fruit more attractive than the flower) which belongs to the same family.

A sun-lover, I like plenty of yellow on the hillsides and in gardens—sunflowers, Californian poppies, winter jasmine, St John's Wort, buttercups, wild strawberries, mustard in bloom... But if you live in a hot place, you might prefer cooling blues and soft purples—forget-me-not, bluebells, cornflowers, lavender. I'd go far for a sprig of sweetly-scented lavender. To many older people the word lavender is charm; it seems to recall the plaintive strain of once familiar music—

Lavender blue, dilly dilly,
Lavender green,
When I am king, dilly dilly,
You'll be my queen.

This tame-looking, blue-green, stiff, sticky and immovable shrub holds as much poetry and romance in its wiry arms as would fill a large book.

Most cultivated flowers were originally wild, and many take their names from the botanists who first 'tamed' them. Thus, the dahlia is named after Mr Dahl, a Swede; the rudbeckia after Rudbeck, a Dutchman; the zinnia after Dr Zinn, a German; and the lobelia after Monsieur Lobel, a Flemish physician. They and others brought to Europe many of the flowers found growing wild in tropical America, Asia and Africa.

But I am no botanist. I prefer to be the butterfly, perfectly happy in going from flower to flower in search of nectar.